Heidegger, Reproductive
Technology, & The Motherless Age

Dana S. Belu

Heidegger, Reproductive Technology, & The Motherless Age

palgrave
macmillan

Dana S. Belu
California State University at Dominguez Hills
Carson, California, USA

ISBN 978-3-319-50605-0   ISBN 978-3-319-50606-7 (eBook)
DOI 10.1007/978-3-319-50606-7

Library of Congress Control Number: 2017930811

© The Editor(s) (if applicable) and The Author(s) 2017
This work is subject to copyright. All rights are solely and exclusively licensed by the Publisher, whether the whole or part of the material is concerned, specifically the rights of translation, reprinting, reuse of illustrations, recitation, broadcasting, reproduction on microfilms or in any other physical way, and transmission or information storage and retrieval, electronic adaptation, computer software, or by similar or dissimilar methodology now known or hereafter developed.
The use of general descriptive names, registered names, trademarks, service marks, etc. in this publication does not imply, even in the absence of a specific statement, that such names are exempt from the relevant protective laws and regulations and therefore free for general use.
The publisher, the authors and the editors are safe to assume that the advice and information in this book are believed to be true and accurate at the date of publication. Neither the publisher nor the authors or the editors give a warranty, express or implied, with respect to the material contained herein or for any errors or omissions that may have been made. The publisher remains neutral with regard to jurisdictional claims in published maps and institutional affiliations.

Cover illustration: Mono Circles © John Rawsterne/patternhead.com

Printed on acid-free paper

This Palgrave Macmillan imprint is published by Springer Nature
The registered company is Springer International Publishing AG
The registered company address is: Gewerbestrasse 11, 6330 Cham, Switzerland

Acknowledgements

Many thanks to friends, family and colleagues who patiently discussed ideas, read chapters and shared feedback on the book. A profound thanks to Daniel Greenspan and Andrew Feenberg for their invaluable comments and insights. Sandra Harding, Ann Garry, Julia Sushytska, Ed and Mary Casey of *Outside Philosophy* have provided helpful notes on Chapter 5. I thank The Center for Subjectivity Research in Copenhagen and The Faculty Legacy Fund at California State University, and Dominguez Hills for their support of my project. I also thank Lissa McCullough, my editor, whose meticulous approach improved this manuscript. I owe a big thanks to my friend and colleague, Jutta Schamp, for her uplifting support. Very special gratitude goes out to my mother for her love and for frequently helping with childcare so that I could write. Finally, my deepest gratitude goes to my amazing son, Piet Stefan, whose birth brought a new sense of energy and commitment to this book.

Contents

1. Introduction: Phenomenology, Feminism, and Reproductive Technology — 1
2. The Paradox of *Ge-stell* — 7
3. Enframing the Womb: A Phenomenological Interpretation of Artificial Conception and Surrogacy in the Motherless Age — 23
4. Mastering the Spark of Life: Between Aristotle & Heidegger on Artificial Conception — 61
5. On the Harnessing of Birth in the Technological Age — 77
6. The *Poiésis* of Birth — 103

Epilogue: Heidegger's *Black Notebooks* — 121

Bibliography — 125

Index — 133

CHAPTER 1

Introduction: Phenomenology, Feminism, and Reproductive Technology

Abstract This introductory chapter provides an overview of the book, including brief summaries of each one of the chapters. It introduces key terms such as enframing, reproductive enframing, phenomenology, feminist phenomenology and identifies some of the central concerns regarding recent advances into reproductive technology.

Keywords Phenomenology · Feminist phenomenology · Enframing · Heidegger · Reproductive Technology

This book engages Martin Heidegger's reflections on the technological age as a way to make sense of the growing use of advanced reproductive technologies (ARTs) in human conception, gestation, and birth. It applies Heidegger's concept of the technological enframing, both critically and appreciatively, to contemporary forms of human reproduction and motherhood, broadly construed. The rising popularity of ARTs is transforming in unprecedented ways the way women conceive, gestate, and give birth as well as the cultural meanings traditionally associated with motherhood, quickly deleting traditional conceptions and experiences of motherhood. For instance, as a result of ARTs it is increasingly common for children to have multiple mothers, even in some rare cases up to four mothers: that is, two biological mothers, one gestational mother, and the social mother. Sexual reproduction and natural (or drug-free) births are

© The Author(s) 2017
D.S. Belu, *Heidegger, Reproductive Technology, & The Motherless Age*,
DOI 10.1007/978-3-319-50606-7_1

increasingly marginalized by high-tech alternatives. Surgical births such as cesarean sections and elective cesarean sections are becoming ever more popular, even fashionable. The latter, especially, bypasses the experience of labor because it extracts the baby without any active input from the mother. But a baby delivered is not a baby born.

As is well known, Heidegger's work on technology belongs to his later philosophical period and his ruminations on the history of being, when he attempts to think the disclosure of being without explicit reference to *Dasein*. In these later writings, as contrasted with his earlier uses of the phenomenological method, Heidegger's phenomenology *historicizes* the understanding of truth in the West as a mode of revealing (*aletheia*) that is presupposed, for example, by the ordinary conception of truth as correspondence. Since Greek antiquity there have been several discrete modes of revealing (*aletheia*), horizons of meaning or epochs, inscribed within an underlying structure of truth as revealing/concealing (*a-letheia*).[1] No single historical epoch or thinker determines that, for instance, truth is "based solely on what Plato thinks as *idea* and Aristotle thinks as *to ti en einai* (that which any particular thing has always been), or what metaphysics in its most varied interpretations thinks as *essentia*,"[2] that is, as *objectivity* for the moderns, as *will to power* for Nietzsche, and as *enframing* for the age to come. The meaning of truth is a function of the historical interplay between the revealing and concealing of different normative horizons that no one controls.

Heidegger's later phenomenology continues to underscore that for something to show up it must show up *as* something. Phenomena are recognizable only within a meaningful context and never when they are viewed abstractly. In his post-World War II essays on technology, this context is historicized so that any particular intentional arc or relationship between human existence and the world is always already circumscribed by a historical framework such as the technological one. Thus, for any set of norms or worlds to be revealed, other norms or worlds must be concealed. These norms vary, but the revealing-concealing structure of being itself within which these variable norms occur is invariable. The enframing is one such variant "upon this overall invariant structure"[3] of being and as such it necessarily conceals other variants.

Heidegger's analysis of the contemporary technological epoch pays little attention to particular technologies and technical users, and much more to the *essence* (*Wesen*) of the technical age and its manner of concealment. He calls this essence *das Ge-stell*, a word that is most often translated as *enframing*. It can be thought to refer to a cultural imprint (*Gestalt*) or to a "mode of revealing" that is specific to our age. As we shall see in greater detail, the phenomenology of enframing describes the relationship between a general attitude of imposition or "challenging-forth," as Heidegger writes in "The Question Concerning Technology," and what this attitude discloses: a world set up as a heap of fungible raw materials, resources, or "standing-reserve" (*Bestand*) awaiting optimization. The dominant value embodied by the attitude of challenging-forth is a constant "driving on to the maximum yield at the minimum expense."[4] This drive places an uncritical value on the control and optimization of resources as the final goal of all action. Enframing does not necessarily reflect the use of technical devices; although devices are often involved, human beings can treat each other and the world as fungible resources without the use of any technical devices at all.

By employing Heidegger's concept of enframing in the specific context of *reproductive enframing*, I interrogate the disclosure of women's reproductive bodies as resources in the enframed stages, for the mother, of medicalized conception, gestation, and birth. This feminist phenomenology can benefit women by raising awareness about the enframing of their reproductive bodies. My feminist phenomenology seeks to provide specific social content to Heidegger's socially anonymous phenomenology while holding on to his fundamental insights. I hope to show that Heidegger's diagnosis of technical culture as enframed is reflected in and confirmed by the current state of ARTs.

The book is intended to be accessible to a wide variety of readers with interests in feminist phenomenology and/or Heidegger's philosophy. Heidegger's views on technology are brought into conversation with such disparate thinkers as Aristotle, Hannah Arendt, Fernand Lamaze, Adrienne Rich, Andrew Feenberg, and Sara Ruddick, among others. The six chapters of the book can be read in continuity with each other, but can also stand alone. Due to the thematic affinity between Chapters 3 and 4 and Chapters 5 and 6, respectively, the reader might find it most profitable to read them as a pair, but this is not necessary. To facilitate a nonlinear reading, each chapter opens with a formulation of Heidegger's concept of enframing that emphasizes different aspects of the theory and its relationship to the specific content of each chapter. The arc of the chapters assumes that the debate over reproductive technology is far from settled. I take seriously the competing (and popular) views that see technology as a tool in our hands with ambiguously good and evil potential uses, and as an anonymous force that sweeps us up into the future, ready or not. I show that Heidegger's theory of technology is closest to the latter view.

The second chapter, "The Paradox of *Ge-stell*," introduces and deconstructs Heidegger's concept of the technological enframing (*Ge-stell*) as a historically inherited attitude in the West, a forgetting of the clearing (*Lichtungsvergessenheit*) that reduces people and things to fungible raw materials or resources. I argue that Heidegger's theory describes the enframed relationship between the individual and her world in two incompatible ways. On the one hand, the individual is completely absorbed by the calculative thinking typical of enframing, so that no alternatives to this calculative mode of being in the world are possible. According to this account, especially emphasized in Heidegger's early essay "Das Ge-Stell" (1949), enframing is totalizing. If this is the case, I counter, then enframing ceases to be problematic. This version that I call *total enframing* would spell the end of history, so that no contrast class, no other epoch, would exist for a meaningful comparison and no thinker could identify enframing as such. On the other hand, if human beings show up as more than, or other than, completely enframed, as Heidegger also insists in "The Question Concerning Technology" (1954), then enframing is incoherent. According to this version, which I call *partial enframing*, calculative thinking cannot be the basis of the intelligibility of this epoch, but must be only one among a number of possible attitudes human beings can take up. Thus, either way—whether as *total* or *partial*—enframing paralyzes itself in attempting to deploy itself. In subsequent chapters, the feminist phenomenological analyses of high-tech conception, surrogacy, and birth reflect the two sides of this paradox and their tension.

In Chapter 3, "Enframing the Womb: A Phenomenological Interpretation of Artificial Conception and Surrogacy in the Motherless Age," I establish the concept of *reproductive enframing* and show how it makes sense of advanced technical interventions into human conception and gestation, especially in vitro fertilization (IVF) and its derivatives, cytoplasmic transfer, testing for male infertility, and the increasingly popular transnational gestational surrogacy. I argue that in view of reproductive enframing the latter is radically reformulating traditional conceptions of motherhood and can be seen to usher in a motherless age. The chapter's main focus is on the medical fragmentation of women's reproductive bodies into parts—eggs, wombs, tubes, hormonal cycles, and so on—that in turn sets them up for continuous medical optimization. This reproductive optimization is distinguished from more traditional processes of objectification and instrumentalization. Reproductive enframing reveals women who undergo IVF, especially gestational surrogates, not as the autonomous agents they are purported to be, nor as merely passive objects, but as available resources in the service of a medical network that seems to run itself.

Chapter 4, "Mastering the Spark of Life: Between Heidegger and Aristotle on Artificial Conception," compares Heideggerian enframing with Aristotle's analysis of the difference between *physis* and *techné* in his *Physics* II. The chapter takes up the question of whether Aristotle's definition of natural entities, as those that have the moving principle or *arché* (of reproduction) within themselves rather than in another, that is, the medical technician (*technités*) or doctor, is applicable to babies conceived and born through advanced medical interventions in the womb, including but not limited to IVF? I argue that with respect to IVF we are unknowingly stuck in a cultural binary, an impasse between Heidegger's and Aristotle's views. On the one hand, we are attached to the power that technology gives us to appropriate or even delete nature; on the other hand, we are attached to the nostalgia of holding on to our nature, a kind of biology that is transmitted whole and intact, despite any and all technical manipulation.

Chapter 5, "On the Harnessing of Birth in the Technological Age," applies *reproductive enframing* to a phenomenological interpretation of modern childbirth. I provide a hermeneutical sketch of the rise of medical intervention into birth from the middle of the nineteenth century in Europe and the U.S. to the present. I argue that this technophilic movement was temporarily interrupted by Lamaze's technophobic movement, which gained popularity during the 1960s and 1970s. This productivist conception of childbirth, initiated by the Soviets and based on the woman as laborer, made progress during the rise of different social movements in the U.S., including feminism. By the late 1970s, however, Lamaze lost ground to a technological conception of childbirth in which women became medical objects. The popularity of ARTs, especially the epidural, contributed to the marginalization of the more woman-centered, Lamazian approach. I argue that both technological and non-technological (but *techné*-centered) approaches to childbirth embody the ontological norms of efficiency and control that Heidegger associates with enframing. Both are enframings (and both are patriarchal), but they are very different in their implications. The enframing is obscured by a deceptive dichotomization, however, that pits natural birth against technological birth. According to my interpretation, the overmedicalized approach enframes by inserting the pregnant woman into the system as an object and later as a resource of technical action. Lamaze enframes by making her the subject of technical action, a kind of worker "laboring" to give birth. These different versions of enframing have feminist implications that Heidegger's view problematically ignores.

The final chapter, "The *Poiésis* of Birth," explores the possibility of nonenframed birth by combining Heidegger's interpretation of *poiesis* as a bringing-forth with Sara Ruddick's concept of maternal thinking as described in her seminal work, *Maternal Thinking: Toward a Politics of Peace*. It turns out that aspects of Ruddick's concept of maternal thinking owe an unacknowledged debt to Heidegger's *Being and Time*, such as her recurring theme that authentic maternal conscience must be a struggle against inauthentic submission to authority. I argue that this struggle promotes self-trust and empathy, values that are integral to a woman's agency and necessary if she is to "bring children forth" rather than having them be "challenged forth." I argue that empathy is necessary for birth. I suggest that the current debate between radical and liberal feminists on informed consent presupposes an empathy-poor world and I advocate for the institutionalization of empathy.

Notes

1. For an excellent account of the difference between *aletheia* and *a-letheia* see John D. Caputo, "Aletheia and the Myth of Being," in *Demythologizing Heidegger* (Bloomington: Indiana University Press, 1993), 9–38.
2. Martin Heidegger, "The Question Concerning Technology," in *The Question Concerning Technology and Other Essays*, trans. William Lovitt (New York: Harper & Row, 1977), 30. See also Patricia Huntington, "General Background: History of the Feminist Reception of Heidegger and a Guide to Heidegger's Thought," in *Feminist Interpretations of Martin Heidegger* ed. Nancy Holland and Patricia Huntington (Pennsylvania State University Press, University Park, 2001), 34.
3. Don Ihde, *Heidegger's Technologies: Postphenomenological Perspectives* (New York: Fordham University Press, 2010), 31.
4. Heidegger, "The Question Concerning Technology," 15.

CHAPTER 2

The Paradox of *Ge-stell*

Abstract This chapter provides a critical interpretation of Heidegger's theory of technology by focusing on its paradoxical dimension. Heidegger claims that the essence of technology or enframing (*Ge-stell*) reduces people, things, and nature to a heap of fungible resources. He wavers between what I call a total enframing, a version that allows for no way out of the enframing, and a partial enframing, a version that allows for the possibility of transcendence. I argue that if humans are *totally enframed* then their essence is compromised and no theory of enframing is conceivable. But if they are only *partially enframed*, then the essence of technology is compromised. The chapter concludes by suggesting how the paradox can be seen to support a feminist phenomenology of reproduction.

Keywords Enframing · Paradox · Ontological · Ontic · Meditative thinking

Martin Heidegger is among the first twentieth-century thinkers to philosophize about modern technology. His writings helped to put the topic of modern technology on the philosophical map, although his engagement with specific technologies never got very far. At first sight, in fact, Heidegger's account of technology appears counterintuitive as the term refers to what may be more broadly construed as the "spirit of the times" (*Zeitgeist*) rather than to a concrete account of specific technical devices. His references to various modern technologies—windmills, airplanes, cars, the lumber industry, the press, radio and TV, clinics, artificial insemination, the human resources department, the mechanized agricultural industry, concentration camps, hydrogen and atom bombs—are introduced and quickly passed over in favor of reflections on what he calls the *essence* (*das Wesen*) of technology. Heidegger's philosophical interest in concrete technologies and in technological culture seems to have been merely incidental; his real interest lay with this essence of technology. In this chapter I offer an overview of his ontology of technology by way of a commentary on its paradoxical nature, and suggest that the paradox may be helpful for a feminist phenomenology of reproductive technology.

Heidegger uses the ordinary German word *Ge-stell*—translated as enframing by William Lovitt decades ago, and more recently as "positionality" and "synthetic compositioning" by Andrew Mitchell and Theodor Kisiel, respectively. Kisiel claims that, "Heidegger's insight into the essence of modern technology as *Ge-stell* proves to be prescient when extended to our own more technically advanced twenty-first century and so provides an illustration of his genius as a philosopher."[1] While I agree with Kisiel's appreciation of *Ge-stell*, I prefer the ease of Lovitt's translation. Enframing describes what Heidegger identifies as the essence of the technological age. The main sources for Heidegger's developed views on the essence of technology are two essays of 1949, entitled "Positionality" and "The Danger," which seem to have been revised and combined to form his now famous essay, "The Question Concerning Technology," published in 1954. In addition to these sources (and the other four related essays collected in The Question Concerning Technology and Other Essays), two speeches of the 1960s—"The Memorial Address" and "Traditional Language, Technological Language"[2]—along with his 1946 essay "What Are Poets For?" underscore and extend the main tenets of his theory of technology.

Heidegger insists that enframing is "nothing technological" (Heidegger 1977, 4) but rather a historical (*geschichtlich*) dispensation, a mode of revealing (*aletheuin*) the being of things as orderable, controllable, and efficient. It is a general attitude of imposition, or challenging-forth (*Herausforderung*) which aims to reduce all things and relationships to mere resources (*Bestand*) awaiting optimization.[3] This technical disclosure is nihilistic because it levels all meaningful differences and hierarchical value systems. According to this technical worldview[4] nature and the world are reduced to fungible raw materials. Occasionally, Heidegger writes as if fungibility also applies to human beings, without qualification. Consider this line of reflection:

> Enframing is, in its setting-up, universal. It concerns everything that presences; everything not just as sum and series, but everything insofar as each entity as such is enframed in its existence as the orderable... Everything that presences in the age of technology does so according to the way of constancy of stock-pieces in standing-reserve. Even the human being presences in this way, even if it seems that his essence and presence is not affected by the setting-up of enframing.[5]

This version of the concept of enframing conforms with a tendency in Heidegger's work to treat the history of being (*Seinsgeschichte*) as a noncausal succession of universal principles of intelligibility that presuppose the forgetting (*Seinsvergessenheit*) of the clearing (*die Lichtung*) as their source. Each epoch(é) is characterized by the way in which beings are given according to such a principle, while *the giving* of the principle— what Heidegger calls the clearing (*die Lichtung*)—is itself forgotten. Epochés are not chronological time periods but, as Thomas Sheehan succinctly puts it, "the awareness of a specific formation of being... while blocking out the clearing that makes that possible" (Sheehan 2015, 257–258). Thus, the technological epoché or enframing is not simply a widespread "problem" we could solve with appropriate remedies, but the underlying structure of being in our time. It is *ontological* rather than *ontic*, to use the terminology Heidegger applied in his earlier work.

The universality of enframing presents us with a paradox, however. According to Heidegger's account of enframing it should be structurally impossible for the enframed subject to know herself as enframed. Heidegger writes, "the challenging Enframing not only conceals a former way of revealing, bringing forth, but it conceals revealing itself," and with it "that wherein unconcealment, i.e., truth, comes to pass" (Heidegger 1977, 27). So, the question arises as to how Heidegger the philosopher could possibly step outside enframing, the universality of which he posits, in order to describe it? If he can do so, the enframing is not universal. But if he cannot, the enframing must remain concealed forever.

In the essay "Das Ge-Stell,"[6] Heidegger presents ordering as a fundamental norm of the technological lifeworld.[7] Its essence is something more than "merely a machination [*Machenschaft*] of people, consummated in the way of exploitation" (GA 79, 29), because, in the technical age people are themselves constrained to order. This constraint is, presumably, most obvious in our handling of machine technology, but is ubiquitous and not restricted to the technical domain:

> This power of ordering allows the supposition that, what is here called "ordering" is not merely a *human* doing, even though the human being belongs to its execution.... Insofar as human representation readily sets up what presences as the orderable in the calculation of ordering, *the human being remains in its essence, whether consciously or not, set up as something to be ordered by ordering.... The human being is ordering's man.... The essence of man is consequently set-up, bringing ordering into human ways.*[8]

Thus, we in the technological age are determined or "set-up" by being as enframing. The truth or unhiddenness (*aletheia*) of technical beings and things remains concealed. "Ordering strikes nature and history, everything that is, and in all ways, how what presences is. What presences is set-up as orderability and is in advance represented as permanence whose stand is determined from out of ordering. What is permanent and constantly present is standing-reserve."[9]

Heidegger's description of this system in these essays is remarkable. Enframing "snatches everything that presences into orderability and is in this way a gathering of this snatching. Enframing is Ensnatching [*Geraff*]."[10] The possibilities of relating to *any and all types* of machine technology are summed up by enframing,[11] yet machine technology is, fundamentally, no mere mechanism (*Räderwerk*), nor is it a particular instantiation of enframing as a universal concept.[12] Rather, enframing is an essential dispensation as that sine qua non required by all machines. "Modern technology is what it is *not only through* the machine, rather the machine is what it is and how it is from out of the essence of technology. One says *nothing about the essence of modern technology* when one represents it as machine technology."[13] The staggering implication is that machine technology is somehow only incidental for understanding the essence of technology.[14]

How does this affect human beings? Heidegger claims that "because man cannot decide, out of himself and by himself, regarding his own essence it follows that the ordering of standing-reserve and enframing *is not only something human*."[15] But insofar as it *is* something human, humans are co-responsible because they exercise a capacity (*Fähigkeit*) for determined participation. The apparent autonomy and self-determination humans enjoy gives the impression that they can opt out of continuous ordering, but this is merely the way that enframing dissimulates itself as the illusion of agency, the way it brackets or blots out its belonging to the hidden clearing. If people "are in their essence already enframed as standing-reserve,"[16] what kind of freedom is this but a mechanical and nihilistic reproduction of the same? When Heidegger insists on the *universal* character of enframing, he underscores this.[17]

In Heidegger's view freedom is to be conceived only ontologically as openness to being in the form of enframing, rather than ontically or instrumentally as the ability of the autonomous agent to choose among a variety of options provided in advance by the social system. Unable to change her urge to order and control, the technicized being is subjected to the imperatives of the system. Substantive goals and meaningful differences between goals are leveled by the ubiquity of technical reason and replaced with a self-optimizing system.

Total enframing thus totally encompasses humans. "To the enframed belongs also man, admittedly in his own way, be it that he serves the machine or that within ordering he designs and constructs the machine. The human being is *in his own way* a stock-piece in the strongest sense of the words *stock* and *piece*."[18] Thus, as technical users and designers, human beings are resources too. Because all activities today are in one way or another technologically mediated, everyone is enframed as either a technical maker, user, and/or designer. The enframed status of the human being would seem to amount to the denial of what is specifically human, however, the historical ability (as receptivity) for disclosing worlds and for grasping this disclosure in thought. Yet this ability is presupposed by the event of enframing itself. The total reduction of everything to raw materials and system components cannot extend to the human being, whose technical way of being is essential to enframing.

How then does the enframing of the technical agent differ from that of the machine? Human beings are neither present at hand nor ready to hand.[19] They cannot be equated with chickens or cows, nor is it convincing to claim that their "own way" of being enframed is simply to serve the system as designers or users of technology; that is, as just another type of system component. Heidegger appears to recognize this:

> The human being is exchangeable within the ordering of standing-reserve. Because he is a stock-piece the assumption holds that he can become the

functionary of an ordering. Nevertheless, man belongs in an entirely different way to enframing than does the machine. This way can become inhuman.[20] The *in*-human is, however, always still in *human*.... The human being of this epoch is, however, enframed by enframing, even when he does not stand immediately in front of machines and operate machinery.[21]

Here human beings are reduced to fungible raw materials—the *in*-human—albeit in a distinctive manner because as in-*human* they are the site of the disclosure of a world reduced to raw material in the first place. This in-human way is not merely unethical or inhumane but indexes an ontological condition bequeathed by enframing as the current configuration of truth. Within that configuration, for Heidegger, the human has a special status of some undefined sort.

It is tempting to read this special status as evidence of a *partial* enframing. But in fact, it is compatible with a total enframing because all the work of enframing happens inside the frame with the peculiarly enframed human being as an essentially passive conduit for the process. Thus, although human beings and things are enframed differently, both *are* enframed. Heidegger is clear that the human being does not stand outside of enframing as its origin or source. This less extreme formulation of the total enframing that accords the human being a special status still raises the reflexive problem of Heidegger's own capacity to understand it. He might be able to at least witness what he is helpless to control. But this should be impossible because an enframed being is by its very nature *only* what it is in the system of operations to which it belongs. Even if the human being is the site of disclosure, it is so in an inhuman way—that is, as enframed—and so it is implausible to attribute to it this type of transcending reflective power. Heidegger seems to agree:

> This is why ordering does not let itself be explained on the basis of any one single case of standing-reserve; it is just as little explainable out of the sum of the previously determined standing-reserves as their floating generality. Ordering does not let itself be explained at all, i.e., it does not lead back to something clear, as something clear that is suddenly given out.... What we care to explain out of this clarity would be entrusted only to thoughtlessness and rash thinking. We are not allowed to want to explain the ordering in which standing-reserve essences (to the extent that explanation leads away from the matter [*Sache*]). We must more so try to experience its unthought essence first of all.[22]

This unthought essence is the epochal nature of enframing as a mode of revealing and a forgetting of the clearing (*Lichtungvergessenheit*). As such, enframing is not identical with the essence of being but is only one of many ways in which being gives itself to *Dasein*. Presumably, if *Dasein* could understand this historical "fact," it would be situated beyond enframing and able to reflect on its situation, as Heidegger himself does. But at the beginning of this passage Heidegger seems to exclude any explanation of enframing. The enframed subject would grasp history too as a resource rather than as a succession of revealings. The enframing is so totalizing that it encompasses the thinker. Heidegger underscores this point when he insists in "Die Gefahr" that "human thinking cannot think the essence of the revealing."[23]

The outcome of the argument for total enframing is paradoxical. Heidegger seems to say that essential though the human being is to the disclosure of an enframed world, no one within that world has the capacity to understand enframing as a historically contingent mode of disclosure, that is, as the true essence of technology. Yet Heidegger does so understand it.

Can the paradox be avoided? Could it be that the human being is only partially enframed or somehow left out of the enframing in instituting it? This appears to be the solution offered in "The Question Concerning Technology," in which Heidegger presents the possibility of a non-enframed point of view situated outside of the discourse it describes. I call this the partial enframing. Partial enframing describes a human being's formal capacity to stand outside of the enframing even as she continues to enframe herself daily and to drive the enframing forward. While this extopian[24] perspective is a necessary condition of the possibility of accounting for the enframing, it simultaneously undermines the universality of the event it sets out to describe as total, all-encompassing for its epoch. In sum, if humans are *totally* enframed, then their essence is compromised and no theory of enframing is conceivable because enframed people by definition cannot reflect ontologically. But if they are only *partially* enframed, then the essence of technology is compromised.

In "The Question Concerning Technology" Heidegger argues that when "man [...] from within unconcealment reveals that which presences, he merely responds to the call of unconcealment" (Heidegger 1977, 19). We cannot help but transform nature into standing-reserve, whether in the guise of protecting and saving the planet or of exploiting it.[25] Environmentalism can be seen as an enframed response to the deterioration of nature because it is an anthropocentric movement in which people arrogate to themselves the right and power to "take care of" or "save" the planet in order to ensure the survival of the human species. While other cultures attribute the protective and sheltering power to the earth itself, technological thinking seeks to reverse this caring through ever more efficient means of ordering and harnessing natural resources, as is evidenced by the sustainability movement. Thus, everything we touch is transformed into a resource. Yet Heidegger adds:

> Precisely because man is challenged more originally than are the energies of nature, i.e., into the process of ordering, he never is transformed into mere standing-reserve. Since man drives technology forward, he takes part in ordering as a way of revealing. But the unconcealment itself, within which ordering unfolds, is never a human handiwork. (Heidegger 1977, 18)

The human being stands in relationship to two levels of being: the unconcealment and the revealed or the real.[26] As claimed by enframing, human beings are the site of *this* historical revealing of being (*aletheia, Unverborgenheit*), that is, the enframing, but they have no control over the structure of revealing as such, the unconcealment (*a-letheia*).[27] No one can choose *not* to belong to *this* revealing; we are simply thrown into it.[28] As in the earlier essays, enframing as a mode of revealing grants intelligibility to the revealed: it is the way in which things make sense for human understanding in modern times.

But just how far is the human being enframed? This is the critical question. Heidegger seems unsure, claiming that the post-modern individual "comes to the very brink of a precipitous fall; that is, he comes to the point where he himself will have to be taken as standing-reserve" (Heidegger 1977, 27). Presumably, standing on the brink is not yet to fall, and so this formulation differs significantly from the earlier claim of total enframing and permits reflective understanding. There are other passages in the later essays that claim that enframing is not yet fully achieved (Heidegger 1977, 33). For instance, in "The Turning" Heidegger writes "When and how it will come to pass after the manner of a destining no one knows. Nor is it necessary that we know. A knowledge of this kind would even be most ruinous for man, because his essence is to be the one who waits."[29]

This reservation opens possibilities that are foreclosed in the earlier essays. We can experience a transformed understanding of a reality to come—presumably by practicing what Heidegger calls essential or meditative thinking (*das besinnliche Denken*). And somehow the mere fact of achieving such an understanding can contribute to the transformation: "If our thinking should succeed in its efforts to go back into the ground of metaphysics, it might well help to bring about a change in human nature, accompanied by a transformation of metaphysics."[30] Meditative thinking does not precede the transformation but is the accomplishment of the transformation. It signals a new way of seeing that Heidegger mostly defines by what it is not. It eschews calculation and the planning of how to maximize output and minimize input. In *Conversations on a Country Path*, Heidegger describes this thinking as a waiting that does not yield knowledge, is not directed toward an object nor toward subjective fulfillment. Rather, this is a "waiting upon" that presupposes a releasement from "transcendental re-presentational thinking"; it allows for a transformation of human nature as "more waitful, more void, apparently emptier, but richer in contingencies" (Heidegger 1966, 74, 79, 83).

Our relationship to technology remains unfree until this transformation comes to pass and we reappropriate our essence as world disclosers, the receivers of new dispensations of being. But this sense of openness to a new dispensation is precisely what is foreclosed by the technological understanding of being. Insofar as enframing is an epochal framework of intelligibility, it simply does not permit the possibilities Heidegger introduces here. We never get clear how this ontological transformation is supposed to be effected or how to achieve "openness" for an epoch "to come." In contemplating this epoch, meditative thinking (*das besinnliche Denken*) presupposes what it predicts and intends to accomplish, namely, the possibility of transcending the enframing.[31]

Still, the preponderance of the text of "The Question Concerning Technology" supports the quasi-transcendental implication that what goes on within enframing cannot sum up the essence of human beings because enframing is disclosed in and through them in the first place. Total enframing threatens the access of *Dasein* to the truth of revealing, but has not yet happened. *Dasein* is "placed between these possibilities" (Heidegger 1977, 26, 28, 32, 33, 35). Thus, as stated earlier in this chapter, in a formal capacity *Dasein* stands outside of the enframing even as one continues to enframe oneself daily and to drive the enframing forward. This describes the partial enframing or the *enframed-enframing double*.[32]

The double, of course, is not the paradox but a way out of the paradox. The paradox cannot be dismissed, although it can be avoided. For instance, in a recent article Lorraine Markotic tries to avoid the paradox by dismissing the claims of total enframing as she argues for an implausible extopian perspective. She refers to capitalism as a totalizing paradigm that cannot be escaped but that can be reflected upon and critiqued:

> Realizing that exchange-value ineluctably affects our relationship to objects (including art objects), *no matter what we prefer or desire*, can make us aware of how thoroughly capitalism permeates us. Although we may not be able to step out of capitalism since it is all-encompassing, we nevertheless may be able to recognize how all-encompassing it is and even describe how it structures thought and experience (in part, because we have learned of precapitalist societies, as Heidegger learned of the ancient Greeks). Similarly, I would argue that even though we are entrenched within Enframing, we can become aware of our entrenchment; to that extent at least, it is not total. (my emphasis)[33]

My point, however, is not that one cannot see or criticize a system within which one is located, but rather that enframing points to a special kind of system that blinds those within it. There is a big difference between market exchange, which has no such blinding power, and enframing, which does. In other words, the social ontology of capitalism permits a reflection that is precluded by the enframing as a revealing, a historically a priori epoch(é) of being, a forgetfulness of the clearing of being. The hermeneutic flexibility entertained by Markotic's analysis is simply not available to the totally enframed individual who, for instance, would interpret the cultures of ancient Greece through the lens of enframing.

At the most general level, this enframed viewpoint is on display in the approach that Western scholarship takes, for instance, to ancient Greek culture, including art, literature, and philosophy. It views this culture as a research repository of information and knowledge to be carefully dug up, archived, annotated, and linked-up, webbed together with other types of knowledge. On a more concrete note, we can see that the figure of Sophocles' Oedipus—an emblem of human reason's horrific encounter with the gods[34]—has essentially withdrawn from view, as today we are in the grip of Freud's analysis of the play. Oedipus has been reframed as a psychological complex whose understanding contributes to a better, more ego-centered organization of our sexual energies, our libido. Recast as a tale about individual libido management, Oedipus reveals the valuation of fungible energy to be directed at will. As such, it can be seen as the drive behind the upholding of monogamy or engaging in criminal sex, building cities, or destroying Europe, as Freud's analysis underlines in *Civilization and Its Discontents*.[35] Hence, the paradox of the two versions of the enframing persists; that is, the fantasy of totality is incompatible with any extopian point of view, and yet the fantasy sustains this point of view. Other commentators have charged total enframing with "fatalism."[36] In my view, however, the paradox, including the total enframing that it presupposes, is neither fatalistic nor optimistic. Prior to any such assessment, it simply reveals two accounts that are inextricably and yet impossibly connected with each other.

In the next chapters, my phenomenological inquiry into reproductive technologies will be framed by the paradox of enframing. In 3 through 5, I take up total enframing in the context of artificial conception and birth. I then contrast this, in the sixth chapter, with possibilities in the arena of childbirth that could reach beyond enframing. I suggest that the paradox of enframing can be useful for a feminist phenomenology of reproductive technology. First, the paradox reveals an authoritative system that appears to dominate directly, by offering no way out, or to dominate indirectly, by eluding knowledge or clarity. As a sending or a gift of being as unconcealment, both versions of enframing refer back to a meta-ontological authority that lies beyond the power of any individual person, group, or sector of society. Second, the paradox reveals a suppressed binary logic of either/or that is atypical of the Heideggerian corpus. Finally, the paradox of enframing can be seen to reflect the paradox of phenomenology itself. In general, phenomenology is necessary to access phenomena that do not show themselves, but rather dissimulate themselves. Whereas in phenomenology what is concealed can be accessed with the right method, the same cannot be said of enframing. As we shall see, the latter conceals itself further and further, especially through the use of various advanced reproductive technologies.

NOTES

1. In his 2012 translation of GA 79, Andrew Mitchell translates *Ge-stell* as positionality. Theodore Kisiel uses the translation "syn-thetic composition-ing." In a recent publication he explains his translation as follows:

    Best translated out of its Greek and Latin roots as 'syn-thetic com-posit [ion]ing,' *Ge-Stell* portends the 21st century globalizations of the internetted WorldWideWeb with its virtual infinity of websites in cyberspace, Global Positioning Systems (GPS), interlocking air traffic control grids, world-embracing weather maps, the 24–27 world programming based on the computerized and ultimately simple Leibnizian binary-digital logic generating an infinite number of combinations of the posit (1) and non-posit (0). The synthetic compositing of computer logic thus maps out the grand artifact of the technological infrastructure that networks the entire globe of our planet Earth.

(See Kisiel's review of Mahon O'Brien, *Heidegger, History and the Holocaust* [2015] in *Notre Dame Philosophical Review* 2016; http://ndpr.nd.edu/news/68003-heidegger-history-and-the-holocaust/).

2. In addition to, what I consider to be, these central sources multiple other writings invoke the essence and dangers of technology. For example, in his 1936 *Contributions to Philosophy: From Enowning*, Heidegger discusses the notion of machination, a precursor to the technological enframing.
3. In part, my account is sympathetic to Peter-Paul Verbeek's critical argument that this ontological story accounts for people and things as mere "outcomes of the history of being." See his *What Things Do: Philosophical Reflections on Technology, Agency, and Design* (University Park: The Pennsylvania State University Press, 2005), 93.
4. Martin Heidegger, "Overcoming Metaphysics" trans. Joan Stambaugh in *The End of Philosophy* (Chicago: The University of Chicago Press, 2003), 104.
5. Martin Heidegger, "Das Ge-Stell," in *Gesamtausgabe*, vol. 79, *Bremer und Freiburger Vorträge* (Frankfurt: Klosterman, 1949), 44. Most translations from GA 79 are mine. This passage reads "Das Ge-Stell ist in seinem Stellen universal. Es geht alles Anwesende an; alles, nicht nur in der Summe und nacheinander, sondern alles, insofern jedes Anwesende als ein solches in seinem Bestehen aus dem Bestellen her gestellt ist.... Alles Anwesende west im Weltalter der Technik an nach der Weise der Beständigkeit des Bestandstücke im Bestand. Auch der Mensch west so an, mag es strecken— und bezirkweise noch so scheinen, als sei sein Wesen und Anwesen vom Stellen des Ge-Stells nicht angegangen."
6. See GA 79. Together with "Die Gefahr," this essay presents an early formulation of the ideas expressed in "The Question Concerning Technology." While "Das Ge-Stell" focuses on the nature of enframing itself, the task of "Die Gefahr" is to inquire into the ontological origins of enframing. It develops a multiplication of ontological levels (the *a-letheic* structure and the enframing as *this* particular *aletheia* or historical dispensation of the deeper structure) to account for technology and emphasizes the possibility for overcoming the enframing through extra-technical means, reflection, poetry, art. In this respect it anticipates the ambiguities of "The Question Concerning Technology" discussed below.

7. See GA 79, 29: "Was ist das Bestellen in sich? Das Stellen hat den Charakter des Herausforderns. Demgemäß wird es ein Herausfördern. Dies geschieht mit der Kohle, den Erzen, dem Rohöl, mit den Strömen und Seen, mit der Luft. Man sagt, die Erde werde hinsichtlich der in ihr geborgenen Stoffe und Kräfte ausgebeutet. Die Ausbeutung aber sei das Tun und Treiben des Menschen." ("What is ordering in itself? The setting up has the character of challenging-forth. Consequently, it becomes a challenging-forth. This happens with the coal, the ore, crude oil, with the rivers and the oceans, with the air. One says that the earth is exploited with regard to its hidden matter and powers. This exploitation, however, is the doing and driving of people.")
8. See GA 79, 30–31: "Diese Gewalt des Bestellens läßt vermuten, daß, was hier >>Bestellen<< genannt wird, kein bloß *menschliches* Tun ist, wenngleich der Mensch zum Vollzug des Bestellens gehört.... Insofern das menschliche Vorstellen bereits das Answesende als das Bestellbare in die Rechnung des Bestellens gestellt hat, *bleibt der Mensch nach seinem Wesen, ob wissentlich oder nicht, für das Bestellen des Bestellbaren in das Bestellen bestellt.... Der Mensch ist... der Angestellte des Bestellens....* Das Wesen des Menschen wird daraufhin gestellt, das Bestellen in menschlicher Weise mitzuvollziehen" (emphasis added).
9. See GA 79, 31: "Das Bestellen betrifft Natur und Geschichte, alles, was ist, und nach allen Weisen, wie das Anwesende ist. Das Anwesende wird als solches auf die Bestellbarkeit hingestellt und so zum voraus als das Ständige vorgestellt, dessen Stand aus dem Bestellen west. Das in solcher Weise Ständige und ständig Anwesende ist der Bestand."
10. See GA 79, 32. This Heideggerian neologism builds on the German verb *raffen*, to snatch or grab for oneself, to hold things together forcefully, to condense; there is no equivalent in English for *Geraff.* I add the prefix *en* to my translation as "ensnatching". Ensnatching denotes the activity of holding things together Holding things together is also reflected in Heidegger's use of the prefix *Ge-* in *Ge-stell* and in *Geraff.* In his 2012 translation of GA 79, A. Mitchell translates *Geraff* as "plundering."
11. Enframing also sums up our relationship to nature so as to constitute the reduction of all of nature to matter and (potential) energy. This underscores a totalizing and post-objective world disclosure. See GA 79, 41–44: "Für die Physik ist die Natur der Bestand von Energie und Materie. Sie sind die Bestandstücke der Natur.... Die Natur ist nicht einmal mehr ein Gegenstand. Sie ist als das Grundstück des Bestandes im Ge-Stell ein Beständiges, dessen Stand und Ständigkeit sich einzig aus dem Bestellen her bestimmt." ("For physics, nature is the resource of energy and matter. They are the stock-pieces of nature.... Nature is no longer an object. She is a fundamental piece of standing-reserve within enframing, a resource whose stand and standing derives solely from ordering"; emphasis added).

12. For a possible explanation of this strange notion, see Feenberg, "The Ontic and the Ontological in Heidegger's Philosophy of Technology: Response to Thomson," *Inquiry* 43 (Winter 2010): 447–448.
13. See GA 79, 34–35: "Die moderne Technik ist, was sie ist, nicht nur durch die Maschine, sondern die Maschine ist nur, was sie ist und wie sie ist, aus dem Wesen der Technik. *Man sagt daher nichts vom Wesen der modernen Technik, wenn man sie als Maschinentechnik vorstellt*" (emphasis added). By analogy Heidegger claims that by means of historical research and calculation it is impossible to determine the essence of history and mathematics, respectively.
14. Peter-Paul Verbeek, *What Things Do: Philosophical Reflections on Technology, Agency, and Design* (University Park: Pennsylvania State University Press, 2005), 92. His analysis underscores this point; "It is not the machines that disclose beings as standing-reserve; rather, the machines exist only because beings are *already present* as standing-reserve."
15. Martin Heidegger, "Das Ge-Stell," in *Gesamtausgabe Band 79 Bremer und Freiburger Vorträge* (Frankfurt: Klosterman, 1949), 38; emphasis added.
16. Ibid.
17. Ibid., 44.
18. See GA 79, 37: "Zum so Gestellten gehört freilich auch, allerdings in seiner Weise, der Mensch, sei es, das er die Maschine bedient, sei es, daß er innherhalb des Bestellens der Maschinerie die Maschine konstruiert und baut. [...] Der Mensch ist *in seiner Weise* Bestand-Stück im strengen Sinn der Wörter Bestand und Stück" (emphasis added).
19. Martin Heidegger, *Being and Time*, trans. Joan Stambaugh (Albany: State University of New York Press, 1996), Chapters 15–18.
20. GA 79, 37, note *p* reads *und ist es geworden* ("and has become in-human").
21. GA 79: 37 reads "Der Mensch ist auswechselbar innerhalb des Bestellens von Bestand. Das er Bestand-Stück ist, bleibt die Voraussetzung dafür, das er Funktionär eines Bestellens werden kann. Gleichwohl gehört der Mensch in einer völlig anderen Weise in das Ge-Stell als die Maschine. Diese Weise kann unmenschlich werden. Das Unmenschliche ist jedoch immer noch unmenschlich.... Der Mensch dieses Weltalters ist aber in das Ge-Stell gestellt, auch wenn er nicht unmittelbar vor Maschinen und im Betrieb einer Maschinerie steht."
22. See GA 79, 31: "Das Bestellen läßt sich überhaupt nicht erklären, d.h. es läßt sich nicht auf jenes Klare zurückführen, als welches Klare wir unversehens all das ausgeben, was uns ohne weiteres und gewöhnlich bekannt ist und gemeinhin als das Fraglose gilt. Was wir aus diesem Klaren her zu erklären pflegen, wird dadurch nur dem Unbedachten und Gedankenlosen überantwortet. Wir dürfen das Bestellen, worin der Bestand west, nicht erklären wollen (h). Wir müssen vielmehr versuchen, sein noch ungedachtes Wesen allererst zu erfahren."

23. Martin Heidegger, "Die Gefahr" in *Gesamtausgabe Band 79 Bremer und Freiburger Vorträge*, (Frankfurt: Vittorio Klosterman, 1949), 50. The passage reads "Menschliches Denken kann [...] an das Wesen der Unverborgenheit [...] nicht denken."
24. Considered as a dystopian narrative, an external or *extopian* observer is both necessary and impossible for the telling of Heidegger's "story".
25. See Oswald Spengler's *Man and Technics: A Contribution to a Philosophy of Life* (New York: Knopf, 1938), 94. Oswald Spengler agrees with Heidegger's analysis of the technical age, albeit without attributing to it an ontological dimension. He writes "We cannot look at a waterfall without mentally turning it into electric power; we cannot survey a countryside full of pasturing cattle without thinking of its exploitation as a source of meat-supply; we cannot look at the beautiful old handwork of an unspoiled primitive people without wishing to replace it by a modern technical process. Our technical thinking must have its actualization, sensible or senseless."
26. In "The Danger" he multiplies the ontological meta-levels to which the essence of technology is indebted. See GA 79, 57: "Die Gefahr verbirgt sich, indem sie sich durch das Ge-Stell verstellt. Dieses selber wiederum verhüllt sich in dem, was es wesen läßt, in der Technik. Daran liegt es auch, dass unser Verhältnis zum Wesen der Technik so seltsam ist. Inwiefern ist es seltsam? Weil das Wesen der Technik nicht als das Ge-Stell und dessen Wesen nicht als die Gefahr und diese nicht als das Seyn selbst ans Licht kommt, deshalb mißdeuten wir gerade jetzt, wo alles doch von technischen Erscheinungen und Wirkungen der Technik mehr und mehr durchsetzt wird, überall noch die Technik. Wir denken über sie entweder zu kurz oder zu voreilig." "The danger hides itself by dissimulating itself as enframing. In its turn this covers itself up through what it allows to be seen, technology. This accounts for our rarely thought relationship to the essence of technology. To what extent is it rare? To the extent that the essence of technology does not appear as enframing and the essence of enframing does not appear as the danger, and the essence of the danger does not appear as Beyng. This accounts for our misunderstanding, in an age traversed by technical appearances and effects, and above all technology. We think about this either too short or too superficially."
27. John D. Caputo, "*Aletheia* and the Myth of Being," in *Demythologizing Heidegger* (Bloomington: Indiana University Press, 1993), 9–38.
28. The later Heidegger introduces sharp and relatively discontinuous historical breaks between the different "modes" of historical interpretation or "revealings" available in the West. See Heidegger, *Contributions to Philosophy (From Enowning)*, trans. P. Emad and K. Maly (Bloomington: Indiana University Press, 1999).

29. Martin Heidegger, "The Turning" in *The Question Concerning Technology and Other Essays*, 41–42.
30. For more on the power of this thinking see Martin Heidegger, "The Way Back into the Ground of Metaphysics" in *Existentialism from Dostoevsky to Sartre*, trans. Kaufmann (Meridian Press, 1975), 267. Also, on the power of meditative thinking and the difference between essential or meditative thinking and calculative or representational thinking see his *Principle of Reason* (Bloomington: Indiana University Press, 1996) and especially his "Memorial Address" and "Conversations on a Country Path" in *Discourse on Thinking* (New York: Harper & Row, 1966).
31. For a detailed account of meditative thinking as constituted by the togetherness of "releasement toward things" (*Gelassenheit*) and openness to the mystery (*die Offenheit für das Geheimnis*) see Martin Heidegger, *Discourse on Thinking*, trans. J. M. Anderson (Harper & Row, 1966), 55. For an excellent account of the possibility of transcending the enframing through the reconstellation of our practices, see Hubert Dreyfus, "Heidegger on Gaining a Free Relation to Technology" (1995).
32. See Michel Foucault, *The Order of Things: An Archaeology of the Human Sciences* (New York: Vintage, 1994), 318–320. The enframed-enframing double is a paradoxical figure of the modern age similar to Foucault's "empirical-transcendental doublet" problematized in *The Order of Things*. Foucault claims that "Man, in the analytic of finitude, is a strange empirico-transcendental doublet, since he is a being such that knowledge will be attained in him of what renders all knowledge possible." Although the constitution of knowledge is not primarily at stake in Heidegger's writings, the impossible convergence between man as the one in charge of the process of enframing and man as the enframed product presents a similar paradox.
33. See Lorraine Markotic, "Paternity, Enframing and a New Revealing: O'Brien's Philosophy of Reproduction and Heidegger's Critique of Technology" *Hypatia*, vol. 31 issue 1, 135.
34. See Sophocles', "Oedipus the King" in *The Complete Plays of Sophocles*, trans. Sir Richard Claverhouse Jebb (New York: Bantam Books, 1982), 77–114.
35. See Sigmund Freud's *Civilization and Its Discontents*, trans. James Strachey (New York: Norton, 1961), Chapters 4–6. I am indebted to Daniel Greenspan for the connection between Heidegger's account of the challenging-forth of physical energy as fungible energy and Freud's account of the organization of the libido as fungible, psychical energy.
36. See Iain Thomson, "Ontotheology," in *The Bloomsbury Companion to Heidegger*, ed. F. Raffoul and Eric Nelson (New York: Bloomsbury, 2005), 327.

CHAPTER 3

Enframing the Womb: A Phenomenological Interpretation of Artificial Conception and Surrogacy in the Motherless Age

Abstract In this chapter I introduce the concept of *reproductive enframing* and show how it makes phenomenological sense of advanced *in vitro fertilization* (IVF) and its various applications, including gestational surrogacy. In view of the *reproductive enframing*, the latter is radically reformulating traditional conceptions of motherhood and ushers in a motherless age. I focus on the medical fragmentation of women's reproductive bodies into parts that in turn sets them up for continuous medical optimization. This reproductive optimization is distinguished from the more traditional processes of objectification and instrumentalization. *Reproductive enframing* reveals women who undergo IVF, especially gestational surrogates, not as the autonomous agents they are purported to be nor as merely passive objects, but as available resources in the service of medical networks.

Keywords Reproductive enframing · In vitro fertilization · Resources · Gestational surrogacy

Heidegger writes virtually nothing about reproductive technology. In his 1954 essay "Overcoming Metaphysics" he makes the following singular remark:

> Since man is the most important raw material, one can reckon with the fact that someday factories will be built for the artificial breeding of human material, based on present-day chemical research...[that]...already

© The Author(s) 2017
D.S. Belu, *Heidegger, Reproductive Technology, & The Motherless Age*,
DOI 10.1007/978-3-319-50606-7_3

opens the possibility of directing the breeding of male and female organisms according to plan and need ... [by way of] ... artificial insemination.[1]

As it turns out more than half a century later, not man but woman—and in particular women's wombs—are becoming the "most important raw material." In this chapter I coin the term *reproductive enframing* to bring out the phenomenological dimension in the "directing" and "planning" or challenging-forth of women's reproductive capacities.[2] Reproductive enframing refers to the putting together (*Ge-stell*) of technical and nontechnical reproductive practices that cast women's reproductive bodies as fungible, disposable, and as targets of self-objectification.

As just presented in Chapter 2, Heidegger claims that the truth about technology has nothing to do with everyday devices used for achieving concrete practical purposes. The essence of technology, according to Heidegger, refers to a prereflective, culturally inherited attitude that frames the world as a heap of fungible raw materials. This attitude displaces modernity's subject-object dualism and introduces a postmodern age, one in which, as Heidegger underlines in his *Conversation on a Country Path*, "the relation between the ego and the object, which [seemed] to be most general, is apparently only an historical variation of the relation of man to the thing" (Heidegger 1966, 78). The objectifying gap between subject and object that is so definitive for modernity's instrumental reason[3] is collapsed in the calculative thinking that is typical of enframing; this is implicit, as I will show, in Heidegger's view. Thus, subject and object are no longer kept apart, no longer "isolated one in terms of the other,"[4] but rather a hermeneutic transformation reveals both as available and fungible resources.

If we recall that the enframing is phenomenologically defined as a relational mode of revealing (*aletheuein*) whereby an attitude of imposition (challenging-forth) discloses and frames people and things as resources (standing-reserve) then we can begin to see how its reproductive variant, *reproductive enframing*, frames women's wombs as reproductive stock (*Bestandstück*). Reproductive enframing includes the following goals: it enhances a woman's reproductive capacities by treating her as a collection of separable reproductive parts; it strives to prevent a woman from feeling pain, especially in childbirth (even when pain does not signal injury or harm); it promises safety and security against a host of detailed potential harms; it expedites conception and birth at the request of the woman and/or at the request of the physician; and it subtends medicalization of the reproduction. Through the lens of feminist phenomenology, this chapter shows how reproductive enframing subtends in vitro fertilization (IVF) and transnational gestational surrogacy, helping to disclose a motherless age.[5]

Reproductive Enframing and IVF

Since the late 1970s, IVF ("fertilization in glass," i.e., in the medical laboratory) has become the gateway medical technology for what Heidegger calls "the organic construction of the human being." Initially developed in the U.S. for the selective breeding of cattle, it has been widely used to successfully fertilize human eggs since 1978, when it yielded its first live human birth, a baby girl named Louise Joy Brown.[6] Since then IVF has helped to produce several million human beings[7] and more are on their way. Initially IVF was marketed to childless, heterosexual, married couples unable to conceive due to infertility. (The medical definition of infertility is "the inability to conceive after 1 year of intercourse without contraception."[8]) Today the market for IVF has significantly expanded and diversified to include individuals who are not affected by infertility.[9] It is used by gay and lesbian couples, by heterosexual couples who want additional biological children, and by women who do not want to experience pregnancy due to health and/or career related pressures, including cosmetic hazards a pregnancy sometimes poses. IVF makes it possible for these persons to have their own genetic children through a gestational surrogate.

While the diversification of users may seem to promote pluralism and to benefit women, as some liberal-feminist commentators on ARTs contend, I will argue that this diversification actually draws the attention *away* from women. According to liberal-feminist theory, IVF increases women's reproductive autonomy by increasing women's control over their options to bear children (before or after a career, with or without gestation). Since choices regarding reproduction continue to loom large for women, increasing the number of reproductive choices would seem to free women up to choose their own paths. But the expansion and diversification of IVF to include gestational surrogates and also egg donors seems to shift the focus *away* from serving women's well-being and toward "making available" women's bodies for enrollment in technologized reproductive systems. Although there is no doubt that IVF benefits some individual women, it does not undermine the patriarchal control of women's bodies *as a group.* In other words, women as a group do not seem to benefit from the expansion and/or diversification of IVF. Moreover, IVF does nothing to empower women who simply do not want to be mothers to stand up for their choice to live a childless life. In fact, it can be seen to tacitly silence this group of women. Today women who are biologically unable to have children seem to be pitied and those who choose not to have children (or postpone pregnancy) due to demanding careers seem to be the object of social contempt. In my view, IVF indirectly exacerbates this oppressive social binary.

Radical and socialist feminists are generally critical of IVF (and of gestational surrogacy in particular) because they see these technologies as advancing patriarchal, classist, racist, and consumerist values. The appeal to reproductive technology, they claim, obscures the social risk factors that lead to infertility and that affect mostly women. These include poor nutrition, poor healthcare, sexually transmitted diseases, and the delaying of childbirth for the sake of attaining professional success in a male-ordered society. IVF works as a band-aid and does not address any of these gender inequities and social pressures. Instead it "places the burden of social problems on women's bodies"[10] and makes it easier for many women not to think critically about the norms and values they identify with, thus reproducing patriarchal norms from which they may derive some individual benefit, but often at the price of furthering oppression of women as a group. Thus, some feminists treat successful IVF results and the well-being of women as distinct issues.

IVF is a medical technology that effectively separates sex from reproduction. As a highly systematic procedure it extracts the egg from the ovary and places it in a Petrie dish where it is fertilized by selected sperm. Prior to this extraction, the woman's reproductive organs are primed for ovulation. She is administered hormonal injections, nasal sprays, or tablets designed to increase the number of follicles that "ripen" during each menstrual cycle. These medications enable doctors to collect and fertilize multiple eggs at once, thereby increasing the chances of achieving pregnancy. Blood tests and ultrasound are used to monitor the growth of the follicles. This requires regular visits to the doctor. When the follicles are "mature" an injection of the female reproductive hormone hCG (human chorionic gonadotropin) is administered. A day or so later the eggs are collected by inserting a fine needle into the ripe follicles and aspirating the oocytes. The needle is passed through the vaginal wall to the follicles using ultrasound guidance. A light anesthetic is administered. Usually several eggs are removed in the hopes that once fertilized in the dish, at least one will become successfully implanted in the womb.[11] While the success rate of the fertilization outside the womb stands at 80 percent, only 10–20 percent of these eggs become viable inside the womb. In order to increase the chances of conception the transfer must occur within two or three days after fertilization.[12] The fertility treatment for superovulation seems to adversely affect the embryo's ability to implant back into the "superovulated" womb. It appears that "an asynchrony of the ovarian and uterine cycles may result from ovulation induction."[13] In an attempt to avoid this asynchrony doctors began experimenting with implantation in a younger womb that had not been superovulated. This fragmentation of reproduction requires an additional woman or women to become a part of the technological reproductive network even as the introduction of additional women further promotes the fragmentation process. Although the participation of more women produced better results[14] the majority of implanted embryos continue to fail to produce a live baby.

While some women are asked to provide the use of their wombs for the sake of improving IVF success rates, others are asked to donate their eggs. Using donor eggs from younger women increases the success rate of IVF in older women and in women whose eggs may have aged prematurely. Young women (ages 18–32) are enticed to donate eggs by the prospect of extra cash and/or reproductive altruism. The disclosure of side effects is minimized. There is no medical follow up for this procedure and no longitudinal study of its impact on women's health. Surprisingly, in the last 30 years, since this procedure has become relatively routine, "not one long term research study has been undertaken to track the health of women who have been egg donors."[15] The women are simply forgotten. This is especially worrisome in light of severe (immediate) side effects associated with egg donation. These include excessive vomiting, fever, stroke, sterility, and ovarian hyperstimulation syndrome (OHSS), which can be fatal. Long-term side effects have been linked with ovarian and breast cancer.[16] Some medical professionals have noted that recurring cases of severe side effects show that society may be comfortable "trading the health of women today for the reproductive health and freedom of women tomorrow."[17] When egg donors are not clearly informed of all known side effects, when post-op care is not available, and when this lack of care and empathy becomes normative, egg donors are treated as fungible carriers of reproductive energy.

On my Heideggerian feminist phenomenology, IVF sets women up as collections of movable reproductive parts or organs. Reproductive enframing sums up the manipulation of the womb's potential by casting it as separable from the woman's body with which it was traditionally regarded as forming a whole. This manipulation introduces a fragmented approach to conception, one that frames the womb as a collection of discrete and movable reproductive parts: ovaries, follicles, eggs, fallopian tubes, hormones, and so on. These parts are managed as stock, potential reproductive energy that is challenged forth for further medical research and experimentation. This vital "energy" is "unlocked," "transformed," and "stored-up" as stock standing by, "on call for a further ordering" or optimization (Heidegger 1977, 16–17). Qua stock, each woman shows up as interchangeable with other women in a medical network that operates without her qua individual person. If the implantation succeeds, then a systematic medical monitoring of the pregnancy begins. If the implantation fails, then the woman is abandoned by the medical industry. Whether surveilled or abandoned, the woman and her embryo are treated as stock for future "ordering" along the "interlocking paths" of science, technology, and economics. A two-step fragmentation process emerges in this medicalized disclosure of a woman's reproductive body. The woman as a feeling and rational subject is reduced to a malfunctioning womb which is then further reduced to a collection of parts to be assessed and optimized. Let us examine this reduction to a resource in more detail.

In "Das Ge-Stell" Heidegger describes the resource or "stock" as follows:

> What the [medical] machine produces, piece by piece, it places in the standing-reserve of the orderable (*Bestellbaren*). The product is stock.... The stockpiece (*Bestandstück*) is something different than the part. The part shares itself with other parts in the whole. It takes part in the whole, and belongs to it. (It completes the whole.) The piece, on the contrary, is separate and is as a piece closed off from other pieces. It never shares itself with these others in a whole. Nor does the resource piece share itself with others like it in standing-reserve. On the contrary, the resource is made piece-meal for orderability.[18]

Moreover, he highlights the fungible character of stock as follows:

> Stock pieces are piece by piece the same. Their stock character demands this uniformity. As the same, the pieces are in extreme competition with each

other; in this way they raise and secure their stock character. The uniformity of the pieces guarantees (*verstattet*) that all pieces are interchangeable on the spot. A stock-piece is replaceable by another. The piece is, as a piece, put up for exchange. Stock-piece means that what is delimited as a piece is exchangeable in the ordering.[19]

Finally, he writes:

Standing-reserve... places in use in order to turn over in ordering. Usage places everything in advance so that what is set up follows what is successful. This is how everything is set-up: as a consequence of... The result (*Folge*) is ordered in advance as success (*Erfolg*).[20]

In pursuit of new clients and research grants, fertility clinics order their success rates in advance, often inflating and conflating the numbers of live births. For instance, clinics repeatedly misrepresent their success rates by reporting IVF success rates as successful *in vivo* implantation and/or live births, which are typically much lower. Moreover, the "IVF pregnancy rate is usually based on the chance of getting pregnant *after* undergoing egg retrieval."[21] While this information is well known to medical professionals, it is not well known to the public. The manipulation of success rates entails a manipulation of the participants, the women who opt for IVF on the basis of misleading statistics.

By deliberately manipulating the statistics, individual fertility clinics[22] knowingly prey on women's most personal and for many their most ardent hope, that is, having a child. This treats the woman merely as a medical resource, a fungible entity, and a means for further research. Her disappointment "need not be taken into account."[23] Her subjectivity is disregarded. I will now turn to philosopher Andrew Feenberg's theory of "primary and secondary instrumentalization" in order to detail women's resource status in standard IVF procedures.

In *Questioning Technology*, Feenberg provides much needed content to Heidegger's theory of enframing, a theory Feenberg critiques as essentialist because he sees it as reducing technologies to their functional dimension. Feenberg's two-level instrumentalization theory refers to the "functional constitution of technical objects and subjects" and to their actual place in the lifeworld as the "realization of the constituted objects and subjects in actual networks and devices."[24] In the chapter "Impure Reason," Feenberg refers to the first and second levels respectively as *primary* and *secondary* instrumentalization. Transposing the elements of his theory to the realm of artificial reproduction helps to critically illuminate the "functional reduction" and fungibility of the woman. Feenberg's theory criticizes the reductive understanding of technology that sees devices merely as *functional* things. He insists that function depends on social context and is of contingent value only; while Western industrialized countries value the function of technologies, other cultures place the emphasis elsewhere. Feenberg writes:

> What differentiates technology and tools in general from other types of objects is the fact that they appear always already split into "primary" and "secondary" qualities, i.e., functional qualities and all others. We do not have to make that distinction deliberately as we would in the case of a natural object since it belongs to the very form of the technical device. Thus an initial abstraction is built into our immediate perception of technologies. That abstraction seems to set us on the path toward an understanding of the nature of technology as such. However, it is important to note that this is an assumption based on the form of objectivity of technology in our society. Function is not necessarily so privileged in other societies. The functional point of view may coexist peacefully with other points of view, religious, aesthetic, none of which are essentialized. (Feenberg 1999, 211)

Feenberg's theory strives to avoid a monolithic and reductive understanding of technology by foregrounding the connection between secondary instrumentalization (or the social integration of new technologies) and primary instrumentalization (or their function). His theory allows us to see how the socialization of IVF underscores the resource status of the woman and her eggs.

In "Impure Reason" Feenberg analyzes primary instrumentalization into four component steps, or what he calls "reifying moments of technical practice." These are decontextualization, reductionism, autonomization, and positioning.[25] I will focus on the first three. A phenomenological interpretation of conception through IVF shows that the lifeworld of the woman as a whole person and potential mother is concealed even *as* she is revealed (to the medical gaze and to herself) as a collection of malfunctioning reproductive parts that need to be fixed. This is where Feenberg's decontextualization of the "object" comes in. He writes "To reconstitute natural objects as technical objects, they must be 'de-worlded,' artificially separated from the context in which they are originally found so as to be integrated into a technical system. The isolation of the object exposes it to a utilitarian evaluation."[26] One might object, however, that the egg is not transformed into a technical object; rather, its potential is actualized through this technological intervention.[27] Whatever the case may be, eggs are often isolated in order to test their reproductive usefulness, as well as the usefulness of the sperm.

Once they are separated from the womb, the eggs "reveal themselves as containing technical schemas, potentials in human action systems which are made available by decontextualization."[28] This means that they are now made available for fertilization, freezing, or to be stored as embryos for future implantation or experimentation. Cryopreservation opens up possibilities for embryo research and experimentation often *unrelated* to reproduction and that may be unknown to the donor. Whether immediately fertilized and implanted or cryopreserved through vitrification, the decontextualization of the eggs reveals the woman and the eggs as stock, fragmented into a collection of *interchangeable* reproductive parts. This is a significant step in the control and ordering of human (re)production and a steppingstone toward more advanced IVF-derived technologies.

Decontextualization is coupled with a second step, *reductionism*, in which the natural object is reduced to its primary qualities, such as "size, weight and shape" or anything else about the "object that offers an affordance."[29] In the case of the eggs, doctors seek high-quality (functional) eggs that contain the proper chromosomes, are young enough and resilient enough to combine with sperm, and energetic enough to divide and multiply after fertilization.[30] The eggs are reduced to these primary qualities because those seem most conducive to technical production: that is, embryo fertilization, growth, and implantation. Whatever the secondary characteristics of the eggs, they remain undiscovered. Feenberg notes "Secondary qualities are what remains, including those dimensions of the object, that may have been most significant in the course of its pretechnical history. The secondary qualities of the object contain its potential for self-development."[31] Feenberg provides the example of a tree whose secondary quality as "habitat" no longer nourishes and shelters numerous species of flora and fauna once it is reduced to its primary quality, that is, a cylinder of wood. It is unclear what the secondary qualities of these extracted eggs may be, but they might relate to the "lifeworld" of conception, an environment that is especially disturbed by superovulation.

Finally, the reproductive enframing in IVF is underscored by what Feenberg calls the process of autonomization. Autonomization refers to the interruption of the *reflexivity* of technical action, its impact on the user, so that the subject can affect the object of technical production seemingly without being affected in return,[32] or being only slightly affected. The autonomization of IVF is clearly visible when the medical staff fails to care for the woman's hurt feelings, mental distress, and/or collapsed life project after an IVF cycle fails. By choosing to dismiss the patient's distress the medical industry promotes an administrative or "purely functional" relationship with its patients. This affords it a kind of immunity from the consequences of its actions and casts the woman as fungible.

In addition to primary instrumentalization, all technical production also involves what Feenberg calls secondary instrumentalization. Since there is no such thing as IVF in-itself, the steps of decontextualization, reduction, and autonomization loosely correspond to moments in secondary instrumentalization—a process that refers to the lifeworld or the social realization of the technology (and that can be distinguished from its primary counterpart only analytically). Secondary instrumentalization involves systematization, mediation, and vocation (Feenberg 1999, 205–206). According to Feenberg, "To function as an actual device, isolated, decontextualized technical objects must be combined with each other and re-embedded in the natural environment. Systematization is the process of making these combinations and connections... of 'enrolling' objects in a network" (Feenberg 1999, 205). In IVF, systematization refers to IVF's commercial and social recontextualization. This means that the fertilized egg, which now appears as a technical object, must be reintroduced in the living womb of a woman and the woman must be successfully integrated in a network of doctor's visits, plus administrative and medical protocols. Since many women who undergo IVF are older, sometimes well into their forties, the social recontextualization of an older pregnant woman challenges traditional values, especially ageist prejudices about conception and motherhood. All of this involves ethical mediation. Feenberg writes "Ethical and aesthetic mediations supply the simplified technical object with new secondary qualities that seamlessly embed it in its new social context.... Recently, medical advances and environmental crises have inspired new interest in the ethical limitations of technical power" (Feenberg 1999, 206). As noted above, the ethical limitations are foregrounded when a woman is deceived about her chances of having a baby and when her distress is medically dismissed after failed IVF cycles. She is treated as disposable, a fungible resource for the technology. In this case, attention to the ethical mediation reveals a *lack* of care for the well-being of the whole woman. This lack arises as a consequence of the overidentification of technology with function, with efficient conception.

Finally, as mentioned in the discussion of primary instrumentalization above, the autonomization process refers to a lack of reflexivity on the part of the doctors and the medical staff. Autonomization corresponds to what Feenberg calls *vocation* in his secondary instrumentalization theory. He characterizes vocation as follows:

> The technical subject appears autonomous only insofar as its actions are considered in isolation from its life process. Taken as a whole, the succession of its acts adds up to a craft, a vocation, a way of life. The subject is just as deeply engaged as the object.... The doer is transformed by its acts.... The rifleman will become a hunter, the worker in wood becomes a carpenter. Vocation is the best term we have for this *reverse impact* of tools on their users. (Feenberg 1999, 206; my emphasis)

So, vocation refers to the subjectification of the technology, its transformative power, whereby through repeated uses the technology is consciously absorbed and made one's own. Over time, this internalization shapes the identity of the user.

What is curious about the use of IVF is that it is widely solicited and yet, when it is successful, its significance is downplayed. In other words, every effort appears to be made—by the media, the medical industry, and even by the women themselves—to frame pregnancy by means of this invasive medical technology as if it had been achieved without the technology; as if the process had no impact on the woman. In the last few years, women have begun to opt for a "mild IVF cycle"—that is, a shorter cycle with fewer shots—because "mild IVF" stays closer to "mother nature."[33] This perception is as preposterous as it is interesting; since even mild IVF relies on disembodied fertilization and acutely medicalizes conception, it is quite removed from whatever "mother nature" might be taken to mean. Invoking "mother nature" has the effect of undermining the role of the technology used for the precise purpose of suppressing "mother nature." This gesture expresses what Kelly Oliver calls, the *mother-effect*, that is, the erasure of the mother figure or of mother nature in order to mythologize it as power and origin (I return to the *mother-effect* later in this chapter).

In general, women for whom IVF works tend to, alongside their doctors, take a technical, abstractly functional attitude toward reproduction. They tend to ignore the impact that the use of IVF technology has on them and their fetuses. Not surprisingly, successful results ameliorate the stress of the procedure, with some women reporting that they "forgot"[34] all about the stress (I return to this point in the next chapter). Since the social identity or the subjectification of the technology is outright denied or suppressed at the individual level, the identity is then supplied by contemporary, advanced industrial societies as the mere consumer of an expensive medical service. Consumers of this service are not necessarily infertile and they do not necessarily desire their own genetic offspring. For instance, IVF is pursued by *fertile* women for the sake of testing the fertility of the woman's male partner. As Françoise Laborie remarks,

> The increasing use of IVF to treat (and diagnose) male infertility means that healthy fertile women are exposed to the dangers of repeated doses of hormones and drugs and major surgeries.... Experiments have been made with what is called 'cross fertilization,' i.e., sperm given by different men are tested for their capacity to fertilize the eggs of a single woman.[35]

This example reveals the reproductive enframing: that is, the reduction of the woman's reproductive body to its function, its ability to serve the interests of men. Thus, it reveals the patriarchal bias at work in "cross fertilization." In fact, this patriarchal bias in the use of IVF is not an isolated occurrence. IVF has enabled preimplantation sex selection that allows doctors to identify the sex of the embryo, thus allowing the future parents to choose which sex they prefer. What was once a matter of chance or nature is now a matter of human choice. In countries such as India and China, preimplantation sex selection (along with amniocentesis) is used to prevent or terminate female embryos and fetuses. According to some radical feminists, such as Dorothy E. Roberts, high-tech reproductive procedures "serve more to help married men produce genetic offspring than to give women greater reproductive freedom.... [They] resolve the male anxiety over ascertaining paternity; by uniting the egg and the sperm outside of the uterus, they allow men for the first time in history, to be absolutely certain that they are the genetic fathers of their future children."[36] Radical feminist activist Gena Corea, for instance, has argued that the final goal of all reproductive technology is to transform women into "mother machines, incubators for life that are controlled by man-made technologies from conception to birth."[37] However, IVF opens up the possibility for a transhumanist future that increasingly dispenses with women altogether; a motherless future of children borne by machines.

Motherless conception through ectogenesis[38]—a term coined by scientist J. B. S. Haldane in 1924—refers to transhumanist conception, or conception outside the womb, with the help of a substitute: a gestating machine, an artificial womb. Once the eggs are decontextualized and reduced to their primary function, they are implanted in a mechanical womb. Ontologically conceived, ectogenesis reflects reproductive enframing by putting together the fragmentation of conception and gestation, the severing of the biological bond between mother and child, and the elimination of women from the processes of pregnancy and birth. The artificial womb can be seen to propagate what Kelly Oliver, following Derrida's deconstructive genealogy, calls the dissemination of "the fable of technology" that "discounts the role of the mother [and] propagates the fantasy of man's triumph over the forces of Mother Nature through which he gives birth to himself without the body of woman/mother. Through *techne* man overcomes both nature and chance by erasing the flesh and blood of the maternal body altogether" (Oliver 2013, 78).

In addition to ectogenesis, fragmentation is essential to advanced, derivative IVF procedures such as intracytoplasmic sperm injection (ICSI) and cytoplasmic transfer. ICSI is a procedure in which a single sperm is physically injected into an egg to cause fertilization: "ICSI is beneficial when there is male factor infertility such as low sperm count or poor sperm quality, [i.e., slow sperm]. It has also been used to increase fertilization rates in older women" (my emphasis).[39] This procedure can be seen to *make* fertilization happen. Rather than merely artificially triggering fertilization by isolating the sperm and egg together in a Petrie dish, ICSI (like ZIFT) seeks to *directly* control the spark of life (I return to this point in Chapter 4). This technology is mediated by a concealed patriarchal bias. Women willingly subject themselves to painful hormone induction and have their healthy eggs extracted in order to serve the reproductive interests of men with deficient sperm.

In recent years, a counterpart procedure to intracytoplasmic transfer, called cytoplasmic transfer (CT), has emerged:

> [It] revitalizes old eggs by combining the nucleus of an older woman's egg (that is, the egg of the woman trying to become pregnant) with the

cytoplasm of a younger woman's egg (that is, the donor). The resulting embryo is thought to be healthier and more likely to implant in the uterus, but it may also contain genetic material from both eggs because the mitochondria in the younger egg's cytoplasm also contain genetic material. (Harwood, *The Infertility Treadmill*, 12)

This procedure reveals the fungibility of the women participating in this process, since each is reduced to her egg-bearing function and her eggs are now cast as "extractable resources."[40] Furthermore, by enabling fertilization in older women, ICSI and CT more or less unwittingly detract attention from feminist concerns with racial and economic gender inequities, such as the lack of support for working mothers, the working poor, and the high demands of career life—inequities that often compel many women to postpone pregnancy until well into their forties when they require IVF.

While IVF enables reproduction with two living genetic mothers, as in cytoplasmic transfer procedures, it also enables the production of offspring with no living genetic mothers at all. This process results in biologically motherless babies, babies whose mothers were never persons: "unborn mothers." In this procedure "viable eggs [are] collected from the ovarian tissue of aborted fetuses for use in fertility treatments such as IVF. Success has been limited; by stimulating the tissue with hormones, researchers are able to develop primary and secondary egg follicles about halfway to the point of maturity."[41] We see how the potential reproductive energy contained in this stock—that is, in the ovarian tissue of the dead fetus—is extracted (or is it stolen?), challenged forth so that, as Heidegger presciently remarked, "the energy concealed in [its] nature is unlocked, what is unlocked is transformed, what is transformed is stored up, what is stored up is, in turn, distributed, and what is distributed is switched about ever anew" (Heidegger 1977, 16). The procedure dispenses with the woman as subject and with the egg as object so that both "disappear into the objectlessness of standing-reserve."[42]

In such cases, the thorny issue of informed medical consent[43] is bypassed altogether since the content of the abortion automatically becomes the property of the medical institution, so that there is no woman to consult. The process dispenses with the need for the female person as biological mother and woman because, as Lisa Guenther points out in *The Gift of the Other: Levinas and the Politics of Reproduction*, the so-called unborn mother is nothing but "a body part without a body, an egg donor but not a person."[44] In fact, there is no "donor" at all and no activity of gift-giving. Rather, the phenomenon is one of extraction, or what Heidegger calls a "plundering" (*Geraff*), as mentioned in Chapter 2. The medical production of "unborn mothers" redefines the meaning of human stock in terms that even Heidegger could not foresee. It introduces a kind of fungibility predicated on fragmentation that was merely implicit in the earlier and more innocuous forms of low-tech reproductive interventions, such as artificial insemination, that still presupposed the presence and cooperation of the woman as person and subject. Here, the subject-object relationship disappears, it is "sucked up into standing-reserve," since the woman as subject is now a body part, that is, viable ovarian tissue, that is, merely an egg *in potentia*, a storehouse of reproductive energy, a fungible resource for a motherless child.

INTERLUDE

Viewed through the lens of reproductive enframing, these advanced forms of technically assisted conception can be seen "not as a 'new technology' but as the end point of a dominant style of thought"[45] that appears to resemble instrumentality but is, in fact, substantially different.[46] In "The Question Concerning Technology" Heidegger insists that the commonplace view that sees technology as an instrument, a means to an end, an object for a subject is "correct but not true" (Heidegger 1977, 6) because it does not capture the *essence* of the coming age of technology. Instrumentality describes a relationship of use between a subject and an object that includes the wrongful objectification of persons and nature, usually for the sake of securing power and/or profit.[47] While instrumental reason is directed toward a specific object and pursues a fixed goal, resource thinking lacks a fixed goal, probes into the dissolution of the object's boundary, is a continuous striving to transform the object into a fungible resource. If there is a final goal, it is a

> will to planetary ordering... [that is] unique in the dimension of its relentlessness and its tendency to colonize everything and everyone... Thus, technological thinking is not just a reflection of the will to master things, it is also the compulsion to master things *on the basis of correct procedure*... [in the continuous pursuit of] better choices, more choices, greater efficiency, a smooth and well-ordered existence.[48]

A remarkable account of resource thinking is provided by Hannah Arendt's analysis of administrative killing in her book *Eichmann in Jerusalem*. According to her interpretation the initial target for the extermination camps was not the Jews. The initial target of this technological extermination in Arendt's account, was disabled Germans. Nazi plans were to expand the technological extermination to include the people of Eastern Europe, whose elimination would produce Hitler's dream of a *volkloser Raum* (a realm or space without people)[49] destined for Aryan Germans. The final target of this high-tech extermination was German "heart and lung patients."[50] Arendt calls this type of orderly and efficient killing "administrative massacre," and her analysis shows that its target group is fungible. Administrative massacre "can be directed against *any given group*, that is . . . the principle of selection is dependent only upon circumstantial factors."[51] Unlike administrative killing, genocide is more specifically object-related, that is, committing atrocities and harboring feelings of hatred against a *specific* group and not another. The latter illustrates instrumental thinking while the former illustrates technological thinking.[52] Enlisted in technological extermination, enframed thinking captures the phenomenon of persecution in reverse, with priority given to the "systematic" extermination of whoever happens to be at hand (*Bestand*). The rigorous emphasis on orderability and efficiency in systematic persecution and extermination distinguishes technological thinking from the less orderly and less protocol-centered instrumental thinking.[53] The latter can be seen to cover up the former.

With respect to reproductive enframing, instrumental reason can be seen to explain the development of IVF and related ARTs for the sake of increasing women's reproductive freedom or for consolidating patriarchal control of women's reproduction by means of ideological choice-talk.[54] On the other hand, the goal of technological thinking is more opaque, less related to the accomplishment of external goals, more invested in the drive to set up new goals and to adopt an attitude of perseverance against all (rational) odds. Carolyn Whitbeck expands this point as follows:

> To regard everything, even ourselves, as a potential resource, is to implicitly regard all possible goals or ends as on a par. As a result, efficiency—that is, the efficient use of resources in the pursuit of goals—is implicitly taken as the overarching value. The determination of goals or ends appears as a matter of personal taste.[55]

In other words, the optimization of all resources becomes the new goal, an end in itself. This optimization and ongoing self-optimization is now an equivalent for happiness rather than merely serving as *the means* for attaining happiness.[56]

In IVF, instrumental thinking fails to explain the drive that keeps women and couples on the infertility treadmill, signing up for more IVF cycles after multiple failures in the hopes of defying odds that are stacking up against them. We notice here a vaguely theological resolution, an element of faith kindred to hoping for a miracle that cuts against rational, object-related instrumental thinking. Instrumental thinking fails to explain the pursuit of a goal with means—that is, reproductive technologies—that do not yield results. Technological thinking, however, is evinced in the resolve that against all odds one's duty is to find and optimize more resources, to never give up. The figure of the gestational surrogate holds up this resolve that paradoxically opens up the motherless world.

Gestational Surrogacy and the Motherless Age

Until recently the sight of a pregnant woman implied impending motherhood. Today, this is no longer the case. The increased use of IVF-based technologies makes it more and more possible that a pregnant woman is simply on the job, working as a maternal surrogate, carrying a baby for someone else. She will not become a mother. Today, a baby's so-called first home is not necessarily the mother's body. In the not so distant future, a baby's "first home" may not necessarily be a woman's body.

Feminists are justified in their concerns that traditional surrogacy promotes patriarchy because it "devalues the mother's biological relationship to the child in order to exalt the father's" (Roberts 1999, 249). This concern is illustrated by the 1986 surrogacy court case between Mary Beth Whitehead and the Sterns, where the biological mother and surrogate, Mary Beth Whitehead was denied custody of her biological child and custody was awarded to the biological father, William Stern and his wife. Although gestational surrogacy does not use the surrogate's eggs and it need not use the intended father's sperm, it continues to devalue the surrogate and her connection to the child. The technological breakthrough is not necessarily a breakthrough for surrogates as a group—a group that lacks unity and political power. Philosophically, the gestational surrogate is an ambiguous and even paradoxical figure. She can be seen at one and the same time as an obstacle to the fulfillment of the fantasy of "man's triumph over nature," the "return of the repressed Mother Nature" (Oliver 2013, 78), and also as the necessary means for advancing technologized birth. In view of reproductive enframing, the surrogates' dual function is conditioned by her resource-being.

When it comes to making phenomenological sense of transnational gestational surrogacy, we might have to overcome what Sandra Bartky aptly called a feminist's "double ontological shock: first the realization that what is really happening is quite different from what appears to be happening, and, second, the frequent inability to tell what is really happening at all." (Bartky 1990, 18) Drawing on accounts from Indian and Israeli gestational surrogates, my Heideggerian interpretation clarifies the resource status of the gestational surrogate and at the same time serves to further disentangle Heidegger's notion of the resource from that of the instrument with which it is often confused. I argue that attempts by surrogates to openly challenge or subvert their instrumental and commodified status—in order to claim a sense of agency—reveal their resource status instead. Extending Heidegger's concept to the phenomenon of gestational surrogacy can do the important work of showing us something that otherwise would not get seen: the surrogate as a resource and the disappearance of motherhood as we have known it. In view of reproductive enframing, the growing practice of transnational gestational surrogacy, also known as reproductive tourism, can be seen to undermine the "whole" mother and to promote a motherless age.[57]

Heidegger typically describes *Ge-stell* using examples taken from industry and nonhuman nature. However in a rare occurrence he connects the lack of maternal care with the essence of the technical age (*Ge-stell*, enframing, positionality). In "The Danger" he writes as follows:

> Objects are no longer permitted much less the thing as thing.... In the essence of positionality the thing remains unguarded as thing.... In our language, where it still inceptually speaks, the word "guard" [*die Wahr*] means protection. In our Swabian dialect this word "guard" means a child entrusted to maternal protection. Positionality...lets the thing go...without the guard of its essence as thing.... In the unguarding of the thing there takes place the refusal of world. (Heidegger 2012, 44–45)

Heidegger swiftly moves past this reference to mothers and children, transposing the absence (or is it a loss?) of maternal care onto the epoch(é), the mode of this technological disclosure itself. Claiming that he is not making a value judgment (*Werturteil*) but articulating a hidden ontological condition, Heidegger refers to (the world of) the thing (*das Ding*) as held in place by a kind of maternal intimacy that suggests a situated historical belonging.[58] This belonging strikes a stark contrast with the world of technically reproducible, fungible system components or stock-pieces (*Bestand-Stücke*) that are enrolled in technical systems (Heidegger 1949, 36–37). Neither thing nor object, the stock-piece becomes the hallmark of the technical age, uprooted and lacking a *place* of its own.

Heidegger's brief reference to the absence of maternal protection leads to a discussion of the (de)worlding of world. When the Swabian dialect speaks "inceptually," it speaks of a world where the child is *being* entrusted to the mother's protection. It does not speak of a world disclosed by the mother's initiative to care for or respond to her child. Rather, she responds to the maternal call of being. In fact, Heidegger does not refer to actual mothers at all but rather to "maternal protection"—a function that presumably could be fulfilled by others, non-mothers. In his short work, *Feldweg*, for instance, Heidegger describes young boys (*Buben*), himself included, playing in his home village, being sheltered by the field path. The field path affords maternal protection (Heidegger 1952, 15). This recurring distinction, albeit implicit, between mothers and "the maternal" can be seen to marginalize mothers from maternal work.

In "The Danger" Heidegger's discussion develops a meta-ontology where the essence of *Ge-stell* is the danger whose essence is, in turn, "beyng" itself (Heidegger 2012, 51). Instead of moving with him, I will linger on this absence of maternal protection and on the worldlessness of the stock-piece or resource. This requires a disaggregation of the concept of *Ge-stell*, bracketing out the unique content[59] of Heidegger's theory: the revealing of entities, especially people, as resources and not merely as instruments or objects. Their resource-being is especially visible in the world of gestational surrogacy, a practice that emerged almost two decades after Heidegger's death, yet emblematic of his theory. Gestational surrogacy holds up the recasting of human beings, that is, women impregnated through IVF, as "pieces of inventory of the standing-reserve" (Heidegger 2012, 34). Their fungible status discloses the coming of a (reproductive) world, one defined by a striving—to borrow Heidegger's words from an interview—to "make the human being, that is, to construct him in his organic being as he is needed, skillful and unskillful (clumsy), clever or dumb."[60] Again Heidegger's thinking seems to show us, without meaning to, the transformation of the phenomenon of motherhood, along with the current understanding of being.

In "Das Ge-Stell" Heidegger claims that to show up as a resource of *any* kind is always already to be ordered, that is, to be "attacked" and "forced into conscription" (Heidegger 2012, 26). Whom the attack benefits or harms, pleases or pains, could be relegated to variable subjective factors, but the fact remains that being a resource at all involves sustaining one kind of "attack" or another. Implicit in this description of violence is a condemnation of violence, a value judgment coiled at the heart of an ontological theory. Heidegger calls *Ge-stell*[61] a "plundering" (*Geraff*) that shows up as a kind of human "exploitation" or drive for "spoils and profit," yet this does not address its *aletheic* or essential dimension (Heidegger 2012, 31, 28).

In "The Question Concerning Technology" Heidegger explains "resource" briefly and in generalities—such as "the way everything presences" (Heidegger 1977, 17) and "whatever stands by in the sense of standing-reserve no longer stands over against us as objects" (Heidegger 1977, 17)—as well as through terse phenomenological examples. In the earlier "Das Ge-Stell," however, his account of the resource is more elaborate, pointing to the (re)ordering of the world into an inventory of stuff that is uniform, equivalent, interchangeable, uprooted, and made available upon request. He writes "The standing reserve is much more that which has been shattered [*Zerstückte*] into the orderable. This shattering does not break apart, but instead precisely creates the standing reserve of the pieces of inventory" (Heidegger 2012, 34). In "Science and Reflection" he claims that the "subject-object relation reaches, for the first time, its pure 'relational,' i.e., ordering, character in which both the subject and the object are sucked up as standing-reserve" (Heidegger 1977, 173). Paradoxically, this "does not mean that the subject-object relation vanishes but rather the opposite: it now attains to its most extreme dominance...a standing reserve to be commanded and set in order" (Heidegger 1977, 173).

Even with his proviso that to order means "to attack and to force into conscription" (Heidegger 2012, 26) rather than to effectively organize, we are left wondering how the resource character of people and things is foregrounded in the dominance of the subject-object relation? Given a dearth of clear explanations it is not surprising that this important phenomenological insight has often been misinterpreted as a lament about instrumentality, objectification, the exploitation of people and nature. Some of his examples, however, go further. They illuminate two features of resources and (at least) two kinds of resources. The two features are the reversal of the traditional subject-object relationship and the absence of pain or discomfort. The two kinds of resources are (1) human resources who order and are ordered by the ordering and (2) nonhuman resources, the merely ordered. I will focus on human resources, who appear to lie both within and without the norms of order, control, and optimization.

Terse phenomenological notes about the tourist industry or the patient supply for a clinic (Heidegger 1977, 18–19), about foresters and radio listeners (Heidegger 2012, 35–36), describe a seamless subordination of subjects to objects where, for example, tourists are ordered to special vacation sites and radio listeners become optimizable fodder for the "public sphere." There is a deeply normative aspect to this ordering "as attack" that paradoxically manifests as lacking coercion and not producing distress. For instance, tourists are not angry at being told what to do. Instead they follow directions well and appreciate their vacations being optimized. Radio listeners enjoy music and news. Most are not concerned about becoming cogs in the public-opinion machine, or that "a requisitioning and positioning...has intervened in the[ir] human essence" (Heidegger 2012, 37). Yet in "What Are Poets For?" Heidegger insists that everywhere a lack of distress is the *real* distress, rendering "the human condition, man's being, tolerable for everybody and happy in all respects" (Heidegger 1971, 116; 2012, 53).

In "the world of technology" constant ordering as "continuous attack and forcing into conscription" discloses a world of uniform, equivalent, and replaceable entities (Heidegger 2012, 42, 62), whose potential is "stored up...distributed...and switched about ever anew," without serving a fixed and final goal, without limit (*peras*) (Heidegger 1977, 16).[62] Almost 10 years later, in "Traditional and Technological Language," he underscores this idea as follows: "What is peculiar to technology resides in...the demand to challenge nature forth into placing it at our disposal and securing it as natural energy. This demand is more powerful than any human positing of ends" (Heidegger 1962, 138). In the social realm, the total mobilization that Heidegger describes can be clearly seen to shape the biopolitics of transnational Indian surrogacy in which the surrogate's reproductive capacities are harnessed and her vital energy is extracted and stored by a network of medical and legal technologies that she also helps to order. She holds up these disparate technologies, their "confluence and interweaving" (Banerjee 2011, 179), without assuming the position of the autonomous subject or falling in the category of exploited object.

My working definition of an autonomous subject refers to someone who thinks rationally, speaks freely with others, and is able to make decisions based on an array of relevant information without suffering from undue constraints. An object is any inert thing with clearly delineated boundaries and functions, separated from other things by spatial and/or temporal intervals. Subjects relate to objects through a process of representation that decontextualizes the object from its immediate environment or its lifeworld. An object is not necessarily fungible, that is, interchangeable with any other object; it has its own standing and purpose. When an object is made continuously available and fungible, interchangeable with other objects, it becomes a resource. Continuous acts of self-objectification turn a subject into a resource. According to Iain Thomson, "once modern subjects dominating an objective world begin treating *themselves* as objects, the subject/object distinction itself is undermined, and the subject is thereby put on the path toward becoming just another resource to be *optimized*, that is, *secured and ordered for the sake of flexible use;* ... self-objectification ... dissolves the subject into the resource pool."[63]

According to my feminist phenomenology the figure of the transnational gestational surrogate discloses a new world of intersecting, medical, economical, juridical, political, and cultural relationships. The burgeoning feminist and ethnographic literature regarding the practice of surrogacy in India leaves little doubt that many of the surrogates are being exploited (Pande 2011; Smerdon 2009; Vora 2013), their human rights violated (Saravanan 2013, 11), and that they are the objects of wrongful instrumentalization and wrongful fungibilization. This is to say they are often treated *merely* as a *means* to an end (Wilkinson 2010, 144–145) and as interchangeable by the clinics, the commissioning couples, and sometimes even by their husbands. I want to turn to another dimension of surrogacy that gets covered up by this discourse of exploitation. It is not the prosurrogacy position supported by many liberal feminists.

In fact, liberal feminists who argue that gestational surrogacy is a woman's free choice and should be respected as such still carry the burden of proof to rescue this "choice" from the constraints of dire poverty and systemic social oppression, including lack of adequate food, lack of housing, lack of education, the burden of drunk and unemployed husbands, and organized "sterilization and aggressive contraception" programs (Bailey 2011, 734), which paradoxically deny Indian women the fulfillment of one of the central roles (motherhood) assigned to them by patriarchy. Cultural anthropologists and ethnographers whose work attempts to move away from *representing* the surrogates in favor of *presenting* their culturally specific narratives argue in favor of the surrogates' *subversive agency*. This agency differs from what Westerners consider reproductive self-determination or autonomy yet, all the same, liberating. As I show below, however, these acts of subversion involve significant moments of self-objectification and self-ordering that reveal the surrogate as a resource, reproducing a network of resources.

In today's market, gestational surrogacy can involve up to four different potential mothers: two genetic mothers, the gestational mother who carries the fetus, and the social mother who provides the care.[64] The gestational surrogate is hired to stand-in, to take the place of another, a future mother who cannot be a mother without this substitution. She agrees to be implanted with the embryo of a couple she does not know and who will claim the baby at birth. This reproductive practice is ethically and politically problematic for the surrogates. According to ethnographer Amrita Pande and feminist philosopher Alison Bailey Indian clinics require that surrogates relocate to dormitories near or in the clinic. Since they must have at least one healthy child to qualify for the program, they must leave their own child/ren behind, motherless for the duration of their surrogacy. They are prohibited from having sexual intercourse with their husbands. Clinics advertise this requirement under the euphemism "dedication."

In addition to cheap prices, surveillance and regulation are selling points for some first world clients, who lament that "that kind of control would just not be possible in the United States" (Bailey 2011, 721). Surrogates are subjected to intense hormone therapy, embryo implantation procedures, and close food and drug monitoring. They are required to sign away all rights to the child and to give permission for cesarean delivery, what Indian surrogates call "the scissors," in advance. Illiterate surrogates sign with a thumbprint (Smerdon 2009, 84). Moreover, a surrogate is not consulted whether or not she agrees to have a multiple pregnancy (Vora 2013, 103). In fact, "the surrogate is mandated to comply with all decisions made about her body" (Pande 2014, 51).[65] Good surrogates will make more money in these 10 months (up to

$10,000) than they would otherwise make working for 9 years crushing glass or cleaning houses. The money will buy a one or two room brick or cinderblock house for her family and may be enough to put her children through school (Bailey 2011, 719; *Wombs for Rent in India*, 2015). The money is not likely to bring gender injustices to a halt, however.[66] In fact, the money that the surrogate earns does not contribute to her individual liberation, education, or transcendence but rather consolidates the grip of Indian patriarchy upon her, keeping her inside the house she purchased.

In the Indian surrogate market the gestational surrogate is known as "a rented womb," the "spare space" (Vora 2013, 99) needed to "house" another woman's embryo, the (genetic) mother. When the embryo belongs to white Western couples, as it does in approximately one-third of all surrogacies,[67] racial "borders are manipulated and transgressed" (Banerjee 2011, 179), but this does not seem to matter. This crossing of racial borders distinguishes the gestational surrogate from the traditional surrogate, who cannot be regarded as merely "spare space" due to her genetic tie to the fetus. Since the gestational surrogate has no such connection, she is more fungible than the traditional surrogate.[68] Her fungibility is exacerbated by her vulnerable social position. In the U.S. and abroad, gestational surrogates are usually minority women and/or poor women who are in dire need of money and the commissioning couples are usually white and/or wealthy.

In the tightly supervised medical world of Indian clinics the gestational surrogate is taught to effectively uncouple from the fetus. The very need for this lesson on detachment reveals that the given or lived experience between the surrogate and the fetus is one of attachment, a bond that must be broken. So, the surrogate is counseled to think of herself as doing God's work, exchanging "life for life" (Vora 2013, 104), and/or being altruistic, helping another woman to become a mother—but she is told never to think of herself as the mother. She is taught how to be affectless and to reject maternal feelings; to disassociate herself from her maternal body, from her lived or personal body (*Leib*),[69] and to see herself as a mere thing, a vessel. This object-body is even more abstract than a vessel, however, since a vessel might be perceived as the appropriate *place* for a thing, as Luce Irigary writes—"a mold that embraces the thing" and gives the thing its being[70]—whereas the language of "spare space" directed at the surrogate reminds her that she is superfluous, inessential to the fetus.

The continuous interruption of the surrogate's connection to the fetus deprives her of the "elemental enjoyment" (Levinas 1969, 163) a pregnancy offers. It teaches her that she is an "I" who is neither mother (the baby has someone else's DNA) nor whore (she is not stigmatized by sexual relations outside of marriage) nor worker (she is not acquiring any skill). According to one doctor, clinics ensure that "no feelings" develop in the surrogate for the child or in the child for the surrogate (Saravanan 2013, 10). The surrogate is seen and taught to see herself as a resource to be commanded and challenged forth by the clinic. Radical feminist Janice Raymond describes the surrogate's fungibility as follows: "She contributes mere...environment, the stock.... As stock, she is...purchased in the manner of a transferable certificate, ... as stock, she is kept for breeding purposes."[71] Once the surrogate begins to see herself as a resource, she often tries to network directly with the commissioning couples for future surrogacies.

All this allows for a glimpse into the resource status of the surrogate. She appears fungible, involved in acts of self-objectification whereby she readily sees *herself* in abstract terms as "empty space"[72] or an "empty container." Thus, as I will further show, she is not simply acted upon or exploited by others, she is engaged in acts of self-ordering and self-objectification even as this "object" becomes increasingly elusive, a resource. As they psychologically separate from the fetus growing inside of them, becoming able to see it as a mere object, a "not-me" that is temporarily housed "within me," some Israeli and Indian gestational surrogates to begin to see themselves as free, fooling nature, and/or subverting the surrogacy system. According to Elly Teman's "Medicalization of 'Nature' in the 'Artificial' Body: Surrogate Motherhood in Israel," the belief that genetic kinship is necessary for motherhood runs so deep with Israeli gestational surrogates that they seem to have no trouble cutting off any relationship with the fetus whose genetic makeup they do not share. They would never consider traditional surrogacy, as they believe this would be tantamount to giving away or selling your own child—a horrible offense against nature and Jewish religious law. As mandated by the state of Israel, moreover, they also refuse to gestate any non-Jewish embryos, such as Arab or Christian ones, thereby patrolling political, religious, and racial borders.

The surrogates also enforce a double subject-object separation between themselves and the fetuses, and between their "natural bodies" (or real bodies) and what they call their "artificial bodies," namely, their pregnant bodies that they perceive to be the result of medical technique alone and that they are convinced are not natural and do not belong to them.

These Israeli surrogates take pride in their belief that they are helping the doctors to fool nature. According to Teman's ethnography, the surrogates' welcoming of medical technology is an act of their *subversive agency*, "willingly relinquishing control of their bodies" (Teman 2003, 86). But this relinquishment is a strange splitting[73] that looks much more like self-objectification rather than subversion. In this splitting the surrogate turns the objectifying gaze—initially directed at the fetus—against herself, reordering her entire body as "artificial," seeing it as separate and other.

The splitting also reaffirms harmful stereotypes about pregnant women (including by the women themselves) as passive vessels—or worse, "spare space"—primarily valued for their reproductive abilities. We can see harmful prejudices about pregnant women finding legitimacy in these gestational-surrogacy ideology and narratives, even though they may be challenged in society at large. The script is ideological because it covers over the myriad ways, emotional and psychological, in which the surrogate is *intertwined*[74] with the fetus, including hormonally through the sharing of her endocrine system. While the surrogate may appear to be "engaged in a long period of waiting where nothing happens" (Young 2001, 55), she actually nurtures the slow growth of a new life—but this is not spoken of. Indian and Israeli surrogates are pregnant but required (and eager) to deny their experiences of pregnancy.

With the help of Heidegger's ontological account of the resource we can see that the typical self/other, subject/object separation[75] actually conceals the resource status of both surrogate and fetus. The surrogate is subordinated to the fetus as the "not-me" by a host of social and medical arrangements. Most surrogates do not feel that this subordination is coercive or painful. They see it as the order of the day, something they must do, "a necessity" (Bailey 2011, 723) or simply "the reality" of the situation (Teman 2003, 86) or the giving of a gift to a "sister" (Pande 2011). While in the West the normative discourse of autonomy is central and surrogates are portrayed as "practical decision-makers," Indian surrogates tend to not speak of it as a choice (Bailey 2011, 722–723). According to multiple feminist ethnographic accounts, many surrogates enter these arrangements hoping to get rehired and to "potentialize their relationships with commissioning couples" for years to come (Vora 2013, 104). In fact, in India gestational surrogacy is beginning to "displace the narrative of exploitation with the discourse of fulfilling unrealized potential" (Vora 2013, 100). Increasingly surrogates see their choice as a way out of poverty and the renting out of the "empty space" in their body makes good sense

(Vora 2013, 100). Local narratives are organized around commercial abstractions such as space and potential. But a self-identification with empty space and/or potential is an identification with *no-thing* at all. It is an identification with anonymous vital energy, fungible and on call, amenable to multiple ends that are continually redetermined. This is the hallmark of the resource.

Some Indian surrogates see themselves as in control of the couple's happiness and calculate how best to benefit from their new relationship. According to one account, surrogate narratives about control and developing a "unique bond with the adoptive parents are a means of coming to terms with their actual *fungibility*" (Panitch 2013, 339; my emphasis). While some might have a degree of control as a *group*, they have very little control as individuals because "the long waiting list of women willing to participate in surrogacy makes interchangeability possible".[76] When doctors at Indian clinics promote surrogacy as an equitable exchange, a "win-win"[77] situation for everyone, this may contradict the exploitation critique of surrogacy, but it supports the phenomenon of surrogates as resources.

We can see that in the surrogate network of commissioning couples, surrogate doctors, artificially conceived embryos, and surrogates for hire, no one player has a clear role as subject or object, everyone is framed as a resource for the network. In "Using Arendt and Heidegger to Consider Feminist Thinking on Women and Reproductive/Infertility Technologies," Maren Klawitter underscores this point:

> The natural ability to bear children becomes part of the standing-reserve.... Contracts are signed that sell the rights to babies who have not even been conceived—babies who, like the potential surrogate mothers, are relegated to the standing-reserve. Everything is secured in advance, carefully controlled and regulated. The medical and business establishments form interlocking partnerships that move buyers (infertile couples) and sellers (potential surrogate mothers) from one to the other. Equally enmeshed is the legal-judicial system which, if called upon, enforces and upholds the interlocking systems. Measures are taken to prevent any woman's attempt, once she has entered into the network, to reject her status as surrogate mother, that is, as standing-reserve.[78]

Thus, hopeful parents-to-be place their order and provide the cash, but otherwise have no real control over the process. The doctors "order" the surrogates in order to maximize the chances of live births for their clients. While the surrogate's body and behavior are ordered to serve the life of the fetus, this fetus itself is a direct product of medical technology. The surrogacy network serves the fetus's survival, but it is not in charge of anything.

Perhaps the surrogate is the object dehumanized by multiple subjects who use their relative positions of power to exploit her mind, her labor, and to deny her agency. Yet reports of exploitation and/or objectification from surrogates themselves seem sparse. In fact, the surrogacy clinics take every precaution to keep the working surrogates qua workers as comfortable as possible, healthy, well fed, well rested, entertained, and well paid (Bailey 2011, 721). While in the past gestational-surrogacy work relied on poor Indian women, in recent years it has expanded to include middle-class Indian women (Bailey 2011, 719).

Combining ethnographic reports and aspects of Bailey's feminist analysis with Heidegger's ontological framework allows us to consider the implications of the contractual relation between transnational gestational surrogates and the institutions that hire them. This commercial relationship reveals them to be neither subject nor object, but resources. The commercial relation ostensibly elevates the surrogate to a level of formal equality applicable to all such relations. In a sense, the legal formality confers upon her a level of subjectivity. So, this is not slavery; she is not an object. Yet material and legal conditions, as Bailey and others show, are so problematic that they evacuate the formal relation of full significance, unable to ensure that her interests are met, encouraging her availability and disposability at the same time.[79] Moreover, since the medical treatment or biological manipulation that the surrogate elects to receive is not for her own sake, as in the case of typical medical treatments, but is for the sake of another—a couple waiting on the outside, usually thousands of miles away—this further undermines her status as subject and points to her resource-being.

Based on Bailey's comprehensive analysis it may seem that the outsourcing couple is the one that mostly benefits from this transaction. The couple's satisfaction reproduces new commissioning cycles. Transnational surrogacy in India is thus not a "third world" phenomenon, but implicates the "first world" that drives it forward, and this global phenomenon is redefining motherhood. Viewed through the lens of reproductive enframing, the *Ge-stell* as it plays out in the arena of transnational gestational surrogacy can be seen to disclose a new world of motherhood.

Heidegger's account of the "disappearance" of the subject and the object into standing-reserve allows us to see the resource status of the gestational surrogate, and with it the emergence of fragmented motherhood, involving up to four[80] women, as the new horizon of contemporary motherhood. The global popularity of Indian surrogacy brings the shift *away* from traditional motherhood into sharp phenomenological focus. Here, by traditional motherhood I mean the notion of the "whole" mother: a woman who conceives, carries, births, and is consistently involved in raising her child (with as much help as she can get). Traditional motherhood is disappearing. Today it is often criticized as "idealized motherhood."

By "idealized motherhood" I am certainly not referring to the universal idealization of mothers everywhere, nor am I claiming that mothers are now, or ever have been, treated well, nor am I necessarily advocating a return to what I call "the whole mother." I do mean, however, that in the age of reproductive enframing the figure of the mother is being replaced by various technologies and *maternal figures* who perform *maternal work*. In this new context, almost anyone[81] can be seen to be a mother—so that no one is *the* mother. In other words, the mother becomes what Kelly Oliver, in her *Technologies of Life and Death*, calls "the mother-effect," or the continuous production of the idea of an original and unified mother that is, however, always already a fiction, her presence never certain. Using Derrida's deconstructive genealogy to blur the binary distinction between (mother) nature and technology, Oliver writes "The *mother-effect*, as we might call it, is the result of the absence of a real mother, who is therefore mythologized and romanticized as the origin and plentitude of Nature, but whose disappearance is a prerequisite for the myth itself."[82] Oliver agrees with Derrida that new reproductive technologies "only highlight what has always been the case" (Oliver 2013, 65), namely, that nature in general and the mother in particular, as the child's unified and certain point of origin, have always been fictions because technology, chance, and secrecy have always been constitutive (yet suppressed) elements of what is taken to be pure nature.[83] I am skeptical about the universality of this claim. The phenomenon of the *mother effect*, especially as it plays out in transnational gestational surrogacy, seems to be unique to the age of reproductive enframing. As we shall see in the next chapters, technological interventions in gestation and in birth, coupled with technological incursions into conception, increasingly edge out the mother's participation from human reproduction while at the same time extolling her significance and worth.

Thus, in a concrete sense, the *mother-effect* describes the disappearance and memorialization of motherhood in a new way, one that hasn't "always been the case."

Notes

1. Martin Heidegger, "Overcoming Metaphysics" in *The End of Philosophy*, trans. Joan Stambaugh (Chicago: University of Chicago Press, 2003), 106.
2. The exclusive reference to women's bodies does not imply that men's reproductive bodies are not enframed but the latter is not the focus of this book.
3. Robyn Ferrell, *Copula: Sexual Technologies, Reproductive Powers* (Albany: State University of New York Press, 2006), 156.
4. Ibid.
5. I would like to introduce a cautionary note about reading Chapters 3 through 6. Throughout these chapters I remain somewhat ambivalent about my ethical position on the proliferation of various forms of ARTs. I realize that this may be difficult for some readers who prefer that I declare whether this technological proliferation and or the technologies themselves are good or bad. While I use feminist phenomenology to bring into critical view the harms and dangers that reproductive technologies pose to women, a final account of specific normative stakes involved in this proliferation belong to a different kind of project.
6. See Karen Dawson's introduction in *Embryo Experimentation* (Cambridge: Cambridge University Press, 1990), xiv–xv. She writes, "Attempts at *in vitro fertilization*... and embryo transfer in animals date back to the late nineteenth century. The procedures were initially developed for the study of maternal effects on the embryo before and after birth, but until the 1950s any claims that IVF had been achieved were met with skepticism. This initial skepticism was justified when it was shown in 1951 that sperm had to undergo 'capacitation' in order to have the ability to fertilize. The 1960s saw IVF and ET established as techniques used widely in animal breeding. Continued research in reproductive biology since then has led to an enormous body of information on the *in vitro* requirements for oocyte (egg) maturation, fertilization and early embryo development for several mammalian species. This knowledge was first applied to the human in 1965.... In 1971 Patrick Steptoe, Robert Edwards and Jean Purdy published a description of the first human blastocyst observed after *in vitro* fertilization. This work was the necessary foundation for the first birth from IVF reported by Steptoe and Edwards in 1978."
7. According to the U.S. Centers for Disease Control and Prevention the other two types of ARTs for human conception are gamete intrafallopian transfer (GIFT) and zygote intrafallopian transfer (ZIFT).

8. Janice Raymond, *Women as Wombs: Reproductive Technologies and the Battle Over Women's Freedom* (San Francisco: Harper, 1993), 3.
9. IVF is used for a variety of nonreproductive medical purposes, research and experimentation.
10. See Karey Harwood, *The Infertility Treadmill: Feminist Ethics, Personal Choice and the Use of Reproductive Technologies* (2007), 26.
11. This medical account can be found at www.lutjenmedical.com.au/ivf-process. Also see Karey Harwood, *The Infertility Treadmill: Feminist Ethics, Personal Choice, and the Use of Reproductive Technologies* (University of North Carolina Press, 2007), 12–14.
12. See Peter Singer and Karen Dawson, "IVF Technology and the Argument from Potential," in *Embryo Experimentation*, ed. Peter Singer, Helga Kuhse, Stephen Buckle, Karen Dawson, and Pascal Kasimba (New York: Cambridge University Press, 1990), 78. This point is underscored by Peter Singer: "If the embryo is to have any prospect of developing into a child, it must be transferred to a woman's uterus. Although the transfer itself is a simple procedure, it is after the transfer that things are most likely to go wrong; for reasons which are not fully understood, with even the most successful IVF teams, the probability of a given embryo which has been transferred to the uterus actually implanting there, and leading to a continuing pregnancy, is always less than 20%, and generally no more than 10%" (ibid.).
13. S. Thatcher and A. DeCherney, "Pregnancy-Inducing Technologies: Biological and Medical Implications" in *Women and New Reproductive Technologies: Medical, Psychosocial, Legal and Ethical Dilemmas* ed. Judith Rodin and Aila Collins (Lawrence Erlbaum Associates, Inc., 1991), 34. For a more detailed account of factors associated with failed embryo implantation, including lost and reticent embryos see Sher, Davis and Stoess, *In Vitro Fertilization: The A.R.T. of Making Babies* (New York: Facts on File, 1995), Chapters 7 and 8.
14. Geoffrey Sher, Virginia Davies and Jean Stoes, *In Vitro Fertilization: The A.R.T. of Making Babies* (New York: Facts on File, 1995), 165.
15. Josephine Quinteville, Founder of *Comment on Reproductive Ethics, UK*. Quoted in *Eggsploitation*: Alexandra, Calla, Kylee, Latoya. Directed by J. Baird and J. Lahl (2010).
16. See Pearson, H. "Health Effects of Egg Donation May Take Decades to Emerge," in *Nature: International Weekly Journal of Science* 442 (August 2006): 607–608. See also "The Possible Association between IVF and Breast Cancer Incidence" in *Annals of Surgical Oncology* 15, no. 4 (April 2008); *Eggsploitation* (2013), directed by J. Baird and J. Lahl, and *Maggie's Story* (2015), directed by J. Lahl. Both documentaries are produced by the Center for Bioethics and Culture Network (CBC) and are intended to raise awareness about dangerous side effects to the egg donors' health.

17. Dr. S. Parisian, former CMO for the Food and Drug Administration. See *Eggsploitation*, dir. J. Lahl (2013).
18. Martin Heidegger, "Das Ge-Stell" in *Gesamtausgabe*, vol. 79 (Frankfurt: Klosterman, 1949), 36; my translation, my emphasis.
19. Ibid.
20. Ibid., 26.
21. Geoffrey Sher, Virginia M. Davis, and Jean Stoess, *In Vitro Fertilization: The A.R.T. of Making Babies* (New York: Facts on File, 1995), 64.
22. Each fertility clinic is supposed to report its results, including live births, to the CDC which then compiles this data and makes it available to the public.
23. Martha C. Nussbaum, "Objectification," in *Sex and Social Justice* (New York: Oxford University Press), 218.
24. Andrew Feenberg, "Impure Reason" in *Questioning Technology* (New York: Routledge, 1999), 203.
25. Ibid.
26. Ibid.
27. The technical origins of in vitro babies will be taken up in more detail in Chapter 4.
28. Ibid.
29. Ibid., 203–204.
30. These standard features are widely available and can also be found online on the popular and sentimental site www.sharedjourney.com
31. Andrew Feenberg, "Impure Reason," in *Questioning Technology* (New York: Routledge, 1999), 204.
32. Ibid., 207–208.
33. See D. Payne, S. Goedeke, S. Balfour, and G. Gudex, "Perspectives of Mild Cycle IVF: A Qualitative Study," *Human Reproduction* 27, no. 1 (2012): 167–172.
34. See M. Verhaak, J. M. J. Smeenk, A. W. M. Evers, J. A. M. Kremer, F. W. Kraaimaat, and D. D. M. Braat, "Women's Emotional Adjustment to IVF: A Systematic Review of 25 Years of Research," *Human Reproduction Update* 13, no. 1 (2007): 27–36.
35. Francoise Laborie, "Looking for Mothers You Only Find Fetuses," in *Made to Order: The Myth of Reproductive and Genetic Progress* ed. Patricia Spallone and D. L. Steinberg (Oxford: Pergamon, 1987), 51.
36. See Dorothy E. Roberts, *Killing the Black Body: Race, Reproduction and the Meaning of Liberty* (New York: Vintage, 1997), 248.
37. Shelley Minden, "Patriarchal Designs: The Genetic Engineering of Human Embryos" in *Test Tube Women* ed. Rita Arditti, Renate Duelin Klein, and Shelley Minden (London: Pandora, 1984), 103. See also Helen E. Longino's "Knowledge, Bodies, and Values: Reproductive Technologies and Their Scientific Context," in Technology & The politics of Knowledge, eds. A. Feenberg & A. Hannay (Bloomington: Indiana University Press, 1995), 198–204.

38. See Shulamith Firestone, *The Dialectic of Sex: The Case for a Feminist Revolution* (New York: Farrar Straus & Giroux, 1970). In the last chapter of the book, the author briefly mentions the artificial womb as the solution to the burdens of pregnancy and birth. This passing suggestion has grown into a nearly full blown feminist, utopian vision about the future of pregnancy and birth. On recent medical progress with artificial wombs, see "Artificial, Womb-Free Births Just Got A Lot More Real" by P. Tadich in *Motherboard* at: http://motherboard.vice.com and https://www.geneticliteracyproject.org/2015/06/12/artificial-wombs-the-coming-era-of-motherless-births.
39. Karey Harwood, *The Infertility Treadmill: Feminist Ethics, Personal Choice, and the Use of Reproductive Technologies* (University of North Carolina Press, 2007), 12.
40. See Thomas Sheehan's *Making Sense of Heidegger* (2015), 258. He writes "But the 'positing' and 'imposition' that Heidegger has in mind with *Gestell* is the particular dispensation that is imposed on us today and that compels us to posit and treat nature and people in terms of *extractable resources*."
41. See Lisa Guenther's *The Gift of the Other: Levinas and the Politics of Reproduction* (Albany: State University of New York Press, 2006), 156.
42. See "The Question Concerning Technology", p. 19.
43. This will be discussed in more detail in later chapters.
44. Ibid.
45. Ibid., 150.
46. See Robyn Ferrell, *Copula: Sexual Technologies, Reproductive Powers* (Albany: State University of New York Press, 2006), Chapter 9. Despite its otherwise insightful interpretation of the Heideggerian *Ge-Stell*, this work identifies instrumental rationality and the subject-object ontology with the ontology of the enframing. Ferrell's account comes closer to Marcuse's understanding of the ideology of technology in advanced capitalism. According to Heidegger, the enframing is precisely covered up by this modern, instrumental ideology.
47. See *One Dimensional Man: Studies in the Ideology of Advanced Industrial Society* (Boston: Beacon Press, 1964). In this book Herbert Marcuse labels the ideology of the instrumental attitude as it is deployed in twentieth-century Western capitalist countries as technological rationality.
48. Louis Wolcher, "The End of Technology: A Polemic," *Washington Law Review* 79 (2004): 341–342.
49. Hannah Arendt, *Eichmann in Jerusalem: A Report on the Banality of Evil*; Revised and Enlarged Edition (New York: Viking, 1964), 107.
50. Ibid., 107, 288. "Between December, 1939, and August, 1941, about 50,000 Germans were killed with carbon-monoxide gas in institutions where death rooms were disguised exactly as they later were in Auschwitz as shower rooms and bathrooms. The gassing in the East—or, to use the language of the Nazis, 'the humane way' of killing '"by granting people a mercy death'—began on almost the very day when the gassing in Germany was stopped."

51. Ibid., 288–289.
52. Marcuse's discussion of technological rationality in *One Dimensional Man* could be seen to explain administrative killing. It provides social and political content to Heidegger's rather abstract conception of technological (calculative) thinking. See Herbert Marcuse, *One Dimensional Man: Studies in the Ideology of Advanced Industrial Society* (Boston: Beacon, 1964), 18, 158, 168–169. One important distinction between these two closely related concepts is that technological rationality emphasizes the responsibilities (and possibilities of transformation) of individual and collective subjects from their current state of ideological reification.
53. In fact, Arendt recounts that when the killing was done in a haphazard and disorganized way, as was the case in Romania, the Nazis were horrified and actually stepped in to save the Jews from being butchered but only so that they may kill them in a more civilized way later (190). It seems that much more than actual hatred of the victim *as* object is at stake. Two kinds of racisms are expressed here, medieval and modern racism. The former describes an attitude of unjustified hatred and persecution of a particular group of people or individuals *qua* members of the persecuted groups. The second type describes an attitude whereby the target group is merely incidental and the drive to dominate through systematization and control is essential. Instrumental thinking reflects the former and technological thinking reflects the latter. It seems clear that, according to Arendt's analysis, the Nazis practiced both technological and instrumental thinking at different times. Unfortunately, the two remain, for the most part undifferentiated from each other and the distinction appears to have been ignored, according to Arendt, by the Israeli court presiding over Eichmann's trial.
54. Notably, Marcuse's in *One Dimensional Man* (1964), Rothman's in *Recreating Motherhood* (1989), and Ferrell's in *Copula* (2006), among others.
55. This does not mean that there are no goals but rather that their meaning is either not broad enough to be culturally shared or that the meaning is broad enough but not fixed, that is, it is shared but only briefly and then quickly shifts (cited in Barbara Katz Rothman's *Recreating Motherhood* [New York: Norton, 2000], 51). In this quote Whitbeck identifies the central features of the enframing but strangely she does not cite it as such, nor does she cite Heidegger. Unlike Heidegger, she interprets the features of the enframing as being a result of capitalist ideology rather than an epochal sending of being in the history of being (*Seinsgeschichte*).
56. I am indebted to Lesley MacAvoy for this insight.
57. See Karey Harwood, *The Infertility Treadmill*: "Gestational surrogacy also challenges traditional views of identity that depend on one mother, one father and a knowable family history through each."

58. This historical belonging seems to hearken back to *Either/Or*, where Søren Kierkegaard speaks of the "motherly love that lulls the troubled one" (145) in Greek tragedy and contrasts it with the rational, ethical calculation of his "present" age. The age in which all is subject in advance to rational calculation and control is one in which the maternal element that lulls the individual—absorbing her into a broader net of forces and histories in which control can't be expected of her, since they are beyond the grasp of reason—must in principle vanish. For more on this distinction see Daniel Greenspan, "Tragedy as Historical Idea: Either/Or's Ancient Drama Reflected in the Modern," in *The Passion of Infinity: Kierkegaard, Aristotle and the Rebirth of Tragedy* (Greenspan: de Gruyter, 2008), 148–149.
59. Here I refer to an *internal* feature of *Ge-stell*. Another unique feature, and one that I consider to be *external*, is discussed by Sheehan in *Making Sense of Heidegger* (2015), 281, as the problematic relationship between *Ge-stell* and the *Seinsgeschichte*.
60. Interview with Martin Heidegger in the mid-1960s, broadcast on a German TV show in 2014 hosted by Joachim Flesh. Here is the full quote in German; the italicized passages are translated in the body of Chapter 3: "Wegen sie diesen Gedanken zitieren mit der Gefährlichkeit der Atombombe und noch [hesitating] einer noch größeren Gefährlichkeit der Technik, so denke ich, an das, was sich heute als Biophysik entwickelt: dass wir in absehbarer Zeit *im Stande sind, den Menschen so zu machen, das heißt, rein in seinem organischen Wesen so zu konstruieren wie man ihn braucht: Geschickte und Ungeschickte, Gescheite und Dumme*. So weit wird es kommen" (my transcription and translation); http://www.dailymotion.com/video/x1tpl46_heidegger-3sat_news.
61. He insists that any attempt to explain *Ge-stell* as a universal concept (*katholou*), its application to technical (ontic) contexts, in an effort to test the limit of "the theory" misses its *aletheic* dimension. We should seek to experience its "unthought essence" instead (Heidegger 2012; [1949], 30). This seems important but difficult. How does one experience an essence? How does one know if one succeeded?
62. One of the defining (and problematic) features of the resource (unlike the object) appears to lie in the interrelated pursuits of continuous availability and the erasure of limits.
63. As Iain Thomson writes, "Once modern subjects dominating an objective world begin treating *themselves* as objects, the subject/object distinction itself is undermined, and the subject is thereby put on the path toward becoming just another resource to be *optimized*, that is, '*secured and ordered for the sake of flexible use*'; self-objectification … dissolves the subject into the resource pool". (Thomson 2005, 60)

64. See Gena Corea, "Egg Snatchers," in *Test Tube Women* ed. Rita Arditti, Renate Duelin Klein, and Shelley Minden (London: Pandora, 1984), 38.
65. They do not receive a contract, so they cannot sue.
66. It may in fact compound these problems, as in cases of postnatal health complications untreated by the clinics, emotional disappointment suffered by surrogates who hoped to further a relationship with the commissioning parents, but who are instead cut off, and, in some instances, children who grow up unable to know the surrogate and who report feelings of anger at being surrogated and living with an incomplete sense of self (see Smerdon 2009, 26). Furthermore, research on the effects of gestational surrogacy on surrogated children is sparse, the effects are undocumented and/or under reported. One critical report is available in the film *Breeders: A Subclass of Women*, directed by J. Lahl. Another report comes from research in Australia. See D. Riggs and C. Due, "Representations of Surrogacy in Submissions to a Parliamentary Inquiry in New South Wales," in *Techne: Research in Philosophy and Technology* 16, no. 1, special issue on Feminism, Autonomy and Reproductive Technology, ed. D. Belu, S. Burrow, and E. Soliday (2012).
67. According to the Australian documentary film *The Baby Makers* (2014), approximately one-third of the 25,000 live births per year at Indian clinics are white newborns. Records are not kept carefully so unfortunately these statistics are not entirely reliable.
68. I agree with Dorothy Robert's claim in *Killing the Black Body: Race Reproduction and The Meaning of Liberty* (1999), that in a racist culture "Feminist opponents of surrogacy miss an important aspect of the practice when they criticize it for treating women as fungible commodities. A Black surrogate is not exchangeable for a white one" (279). While this may be true of a traditional surrogate it does not apply to gestational surrogacy. A Black gestational surrogate is exchangeable for a white one and a white one for a Black one. Both are revealed as reproductive stock and both are able to optimize their reproductive bodies. In fact, as I show in my discussion of reproductive tourism in India, racism and classism increase the chances that disenfranchised women will be selected for gestational surrogacy. Furthermore, the market of gestational surrogates is expanding to other third world countries where women suffer from abject poverty: Thailand, and more recently Mexico.
69. For the distinction between the lived body (*Leib*) and the physical or objective body (*Körper*) see Husserl's *Ideas II*, especially 159, 166.
70. See Luce Irigaray, "Place, Interval: A Reading of Aristotle, *Physics IV*," in *An Ethics of Sexual Difference*, trans. C. Burke and G. C. Gill (Ithaca: Cornell University Press, 1993).
71. See Raymond's "New Definitions of Motherhood and Fatherhood" in *Women as Wombs* (San Francisco: Harper, 1993), 31. Raymond's radical feminist analysis underscores the sexism inherent in all ARTs.

72. "The reformulation of the surrogates' bodies as empty spaces that can be cultivated to reproduce Western society and Western lives recapitulates the colonial epistemology of land as property, where resources, including native labor, were used to sustain the metropole" (Kalindi Vora, S&F Online, 5).
73. This sense of splitting differs in important respects from the compelling psychoanalytical account of splitting (in pregnancy) presented by Julia Kristeva in her well-known essay "Women's Time," in *The Kristeva Reader* (New York: Columbia University Press, 1986). She writes "Pregnancy seems to be experienced as the radical ordeal of the splitting of the subject: redoubling up of the body, separation and coexistence of the self and of an other, of nature and consciousness, of physiology and speech. This fundamental challenge to identity is then accompanied by a fantasy of totality—narcissistic completeness—a sort of instituted socialized, natural psychosis" (206).
74. For more on pregnant embodiment, time, and desire, see Kelly Oliver, "Motherhood, Sexuality and Pregnant Embodiment: Twenty-Five Years of Gestation," *Hypatia* 25, no. 4 (2010): 769–776.
75. This separation is not the blurring of the subject-object relationship described by Merleau-Ponty in *The Visible and the Invisible* (1968) in his concept of the *chiasma*, as a subject-object intertwining or ambiguation.
76. See Saravanan (2013).
77. The expression "win-win" is used by Dr. Nayna Patel in *Google Baby* (2012), an influential documentary film about transnational gestational surrogacy in India.
78. Maren Klawitter, "Using Arendt and Heidegger to Consider Feminist Thinking on Women and Reproductive/Infertility Technologies," *Hypatia* 5, no. 3 (1990): 73.
79. See also Wendy Lee, "Reproductive Technology and the Global Exploitation of Women's Sexuality," in *Contemporary Feminist Theory and Activism: Six Global Issues* (Toronto: Broadview, 2010), 61. Lee's feminist analysis underscores the stock status of the surrogate in India by stating that the surrogate is "in effect simply the machinery through which the product is manufactured; that both machinery and product are living beings is irrelevant to the substance of the transaction, since a failure to meet the conditions of the contract is likely attended by consequences identical to the failure of any contract—however much more seriously they are felt by the surrogate."
80. Up to two women can be involved in the genetic conception of the child; one is involved physiologically, that is, the gestational surrogate, and one is the social mother who may be completely unrelated, genetically and physiologically, to the baby. The presence of two genetic mothers is a result of "mitochondrial donation," that is, transplanting the nucleus from an egg cell with

damaged mitochondria into another egg with healthy mitochondria. The second egg is from a donor. The (small) gene pool in the donor's mitochondria, in this process, becomes part of the baby's genetic makeup. Mitochondrial donation has recently become legal in the UK. See "A Dad and Two Moms," in *The Economist*, 2/7/ 2015. http://www.economist.com/news/britain.

81. This point of view is central to fertility organizations such as *Family by Design*, and others, that maintain that the modern family can be anything one wants it to be. Consequently, the figure of the mother is replaced by maternal work. This maternal work, then, can be carried out by surrogates, au pairs, adoptive mothers and fathers or almost anyone.

82. Kelly Oliver, *Technologies of Life and Death: From Cloning to Capital Punishment* (New York: Fordham University Press, 2013), 57.

83. According to Derrida's deconstructive analysis, in *Of Grammatology*, Jean-Jacques Rousseau's relationship to his dead mother, a mother he never knew because she died, universalizes a situation that alas, may have been rather unique to Jean-Jacques Rousseau. It is plausible that, as Derrida argues, for Rousseau the name Mamma (referring to Madame de Warens), designates a chain of supplements, a chain of substitutes (that includes his life partner, Thérèse) and that relies on a fictitious origin, that is, his biological mother. Derrida writes

> Jean-Jacques could thus look for a supplement to Thérèse only on one condition: that the system of supplementarity in general be already open in its possibility, that the play of substitutions be already operative for a long time and that in a certain way Thérèse herself be already a supplement. As Mamma was already the supplement of an unknown mother, and as the 'true mother' herself, at whom the known 'psychoanalyses' of the case of Jean-Jacques Rousseau stop, was also in a certain way a supplement, from the first trace, and even if she had not 'truly' died in giving birth (156).

According to Oliver, Derrida's psychoanalytic deconstruction shows that Rousseau's political philosophy, that advocates the primacy of nature, is also a surrogate, a substitute for his dead mother. While this may apply to Rousseau and others, it's not clear why this chain of supplements (and what Oliver names "*the mother-effect*") should be taken to apply universally, from the dawn of Western philosophy until today.

CHAPTER 4

Mastering the Spark of Life: Between Aristotle & Heidegger on Artificial Conception

Abstract This chapter enlists the Heideggerian concept of enframing along with Aristotle's distinction between nature and art, in order to attempt to discern where control over the spark of life lies in human reproduction through *in vitro fertilization* (IVF): with nature or with medical technology? I argue that if we continue to think of IVF technologies as giving us a piece of nature, as we currently do, the truly technological and revolutionary dimensions of the machinery and of the cultural outlook remain submerged and invisible. Finally, the chapter questions the patriarchal attachment to parenthood *as* biological ownership that is presupposed by the proliferation of human conception through IVF.

Keywords Nature · Techné · Reproductive enframing · In vitro fertilization · Biological ownership

This chapter enlists the Heideggerian concept of *enframing* along with Aristotle's distinction between nature (*physis*) and art (*techné*) in his *Physics* II, in order to attempt to discern where control over the spark of life lies in human artificial reproduction: with nature or with medical technology? I begin by asking whether Aristotle's account of natural entities as those that have the moving principle within themselves rather than in another (Aristotle 1941, 199b15) is applicable to human offspring conceived through advanced technological intervention in the womb, including but not limited to in vitro fertilization (IVF). By drawing on Aristotle's account of nature as self-moving *enformed* matter (Aristotle 1941, 193a28–31), I bring out the significance of the fertility doctor whose medical outlook and technology push against the limit of nature so far that the line between human reproduction and human production begins to disappear.

As discussed in the previous chapters, Heidegger defines the technological age by the imperatives of order, control, and optimization (Heidegger 1977, 16). The drive to order recasts nature as a "coherence of forces calculable in advance" (Heidegger 1977, 21), an approach that overrides the limit of natural entities. In view of this drive, *techné* (as fabrication), especially in the case of IVF, no longer "partly completes what nature cannot bring to a finish" (Aristotle 1941, 199a16–17), as Aristotle would have it and as today's commonsense holds. Rather, IVF produces what nature stubbornly refuses to conceive. This chapter argues that today we are caught between Aristotle's and Heidegger's views on nature and technology. On the one hand, we are attached to the power technology gives us to appropriate, supplant, and perhaps, delete nature, while on the other hand we are attached to the nostalgia of holding on to our nature, a kind of biology that is transmitted whole and unscathed, despite any and all technical manipulations.

ON THE *LIMIT* IN ARISTOTLE'S ACCOUNT OF NATURAL AND TECHNICAL PROCESSES

The notion that there is a fundamental ontological difference between artifacts and natural things is still prevalent today. We distinguish natural things from technological or artificial things and we like to keep them apart. Out "in nature" we seek to encounter a pristine and untouched environment, free from technical construction, even as the use of technology shapes and mediates experiences we like to imagine are "natural"; that is, self-generated rather than externally produced. Hikes in the wilderness, for instance, are facilitated by meteorological forecasts, technological transportation, and gear that keeps us insulated from the cold and safe from the heat. In general, technology organizes the wildness of nature according to our needs and preferences today more extensively than at any time in the past. My purpose is not to highlight the elusiveness of natural experiences in the age of technology but to have a philosophical look at our collective attachment to the category of "nature" in an unlikely context: that of artificial human reproduction, especially through IVF. Although IVF children are at least in part technologically created, it is a commonplace to view them as completely natural. This view is so deeply rooted that it does not even occur to us for a moment to think about the final product of this invasive technological procedure as remotely nonnatural. Doing so would seem offensive, a little taboo. When it comes to IVF, it is almost as if the triumph of technology demands that the technology itself be disavowed and called nature. But, where exactly is this nature?

In *Physics,* book 2, Aristotle writes:

> Each [natural thing] has *within itself* a principle of motion and stationariness (in respect of place, or of growth and decrease, or by way of alteration.) On the other hand, a bed and a coat and anything else of that sort, *qua* receiving these designations—i.e., in so far as they are products of art—have no innate impulse to change. (Aristotle 1941, 192b14–17)

An artifact does not have "in itself *the source* of its own production" (192b28–29; my emphasis). It is generated by an external cause, that is, the technician, and so the movement from its potentiality to its actuality is not of itself; it is introduced by another. Aristotle underscores this point in his *Nicomachean Ethics,* book 6, in which he writes that, regarding artificial objects, "the origin is in the maker and not in the thing made; for art is concerned neither with things that are, or come into being, by necessity, nor with things that do so in accordance with nature (since these have their origin in themselves)" (Aristotle 1941, 1140a14–17).

Throughout *Physics,* book 2, Aristotle equates nature or the natural with self-initiated change. In preparing his discussion of the four kinds of explanation of movement, for instance, he writes, "this then is one account of 'nature,' namely that it is the immediate material substratum of things which have in themselves a principle of motion or change." He immediately adds a second definition of nature that will be decisive, that is, "nature is shape or form." He explains:

> The form is 'nature' rather than the matter; for a thing is more properly said to be what it is when it has attained to fulfilment than when it exists potentially. Again man is born from man, but not bed from bed...the shape of man is his nature. For man is born from man. (193b7–12)

Natural beings are always already enformed, so matter and form are inseparable *in being* and separable only analytically or "in statement" (193b3).

It is important to stress this point as it will help us to avoid the mistake of identifying nature with either matter (i.e., prime matter) *or* with form, exclusively. Although Aristotle subordinates matter to form—associating the former with passivity and the latter with activity—he underscores that natural entities are the coupling of the two.[1] In his *Generation of Animals* he applies this metaphysics to human reproduction as follows: "If, then, the male stands for the effective and active, and the female, considered as female, for the passive, it follows that what the female would contribute to the semen of the male would not be semen but material for the semen to work upon" (729a29–a30). As Carol Bigwood argues in *Earth Muse*, by subordinating the woman's menstrual blood to the man's semen, Aristotle's theory diverts the contribution of the female element from the embryo itself and subordinates it to the male element, effectively underscoring the activity, and hence the superiority of the latter (Bigwood 1995, 162). By appropriating the maternal element (and the material cause), the father takes all the credit for the offspring, according to Bigwood. Despite this subordination (to the semen/form), and pace Bigwood's interpretation, however, the appropriated matter matters quite a bit.[2] Aristotle claims that matter "is a relative term; to each form there corresponds a *special matter*" (Aristotle 1941, 194b10; my emphasis), suggesting that the matter is somehow, albeit not yet fully, differentiated prior to being enformed. But this matter is not lying around; it is not available for inspection. This point is brought out by Trish Glazebrook in her account of Aristotle on *physis* and *techné*: "Aristotle's universe has nowhere in it unformed matter waiting for some form to be imposed upon it. Rather, generation is a special case of motion. It is a transition of substance to substance. Formed matter becomes differently formed matter" (Glazebrook 2000, 103).[3]

While Aristotle identifies nature with what he calls the active, form-giving, masculine principle rather than with the matter or the passive, feminine principle (193b7–9) he emphasizes that it is "the combination of the two" (193b6), "their primary and appropriate belonging together" (Glazebrook 2000, 107) that constitutes natural beings. We could say that while the limit, form, or boundary is inscribed on/in the matter, the matter is co-responsible for receiving a particular formation. A particular "special matter" responds to the seed (already genetically enformed) that will grow into, say, a rose instead of a tulip. Aristotle adds:

> Those things are nature which, by a *continuous* movement *originated from an internal principle*, arrive at some completion; *the same completion is not*

reached from every principle; nor any chance completion, but always the tendency in each is towards the same end, if there is no impediment. (199b15–18; my emphasis)

Thus, "in natural products the sequence is *invariable*, if there is no impediment" (199b26; my emphasis). This sequence aims at a final end (*telos*) and is drawn out, pulled by this end itself, to itself. As a consequence of imitating nature, human making or *techné* is also teleological.[4] However, nature teaches and is exemplary for human making in a general way only, in terms of "technological purposiveness" rather than in the copying of specifics. This idea is brought out by Joachim Schummer in "Aristotle on Technology and Nature" as follows: "Since the teleology of nature is more stringent than human rationality, the former is exemplary for the latter. Without the exemplary teleology of nature there would not be any human purposive activity and hence no technology" (Schummer 2001, 114).[5]

Technology partly imitates natural teleology on the general level of directivity and purposiveness (Schummer 2001, 114) and "partly completes what nature cannot bring to a finish" (Aristotle 1941, 199a15) through particular innovations and techniques of its own. This implies dual aspects of *techné*. The first aspect refers to a knowing that reveals the whole or *eidos* of a thing as this is delimited by the *telos*, the boundary, limit or end that gives each thing its being.[6] In light of this knowing, particular techniques or ways of making become relevant. When a *techné* is said to complete the work of nature (in cases where there is a parallel between growth and making), it employs various techniques to reach its end. This second aspect of *techné* is invoked in the application of manual skill, in the fabrication of the thing. Sometimes *techné* achieves a different kind of completion than would have been achieved by nature alone. If the statement "nature *cannot* bring to a finish" refers to internal and/or external obstacles that dirempted the flow of nature, then *techné* attempts to restore this flow. For instance, various medications help to restore health. Along similar lines, the use of prosthetics restores, at least in part, a range of movement obstructed by the loss of limbs. Despite this medical success, however, the prosthetics facilitate a somewhat stilted movement compared with the natural kind. Thus, the *telos* afforded by technical devices mimics yet also differs from the *telos* afforded by nature. As we shall see, IVF technologies may produce a baby but they do not restore fertility. Along Aristotelian lines, they can be seen to make the enformed reproductive matter "serviceable."

Aristotle writes that "in the products of nature the matter is there all along" as the "necessary" (200a30) element, while in the products of art we make the material and/or "make it serviceable" (194a33–34). The matter in natural processes is given insofar as it is continually shaped by the form, by nature (Glazebrook 2000, 103). In technical processes, the matter is available to the maker based on his or her contact with social practices, culture, and tradition. The form does not become haphazardly determined in accordance with arbitrary individual tastes or preferences. Rather a socially objective *logos* continues to guide the production process, helping to translate the *eidos* (available in the maker's mind) into a finished thing.[7] For instance, the ubiquitous (production of) amphorae in antiquity required that the potter choose the appropriate clay for purification and then, either by hand or with the help of a pottery wheel, make the culturally normative neck shape, apply the decorative designs and colors that were popular during various stages of Greek antiquity. The selection of the right clay and the purification process make the material serviceable for production. In a similar way, IVF doctors select the appropriate eggs and sperm and apply the correct techniques that alone make what can be seen as the reproductive material serviceable. In other words, they harness its potential energy[8] to produce a child, seemingly more and more from the ground up.

Fertility doctors have become the craftsmen of human life, engaging in the dual aspects of *techné*. They imitate nature in its general purposiveness by striving to secure reproduction. In their choice of technique, however, they cut against nature. For instance, as I discuss in Chapter 3, the woman's potential reproductive energy is unlocked through superovulation, a process that produces multiple eggs (up to ten or more) instead of just one. The eggs are then sucked out of the woman's ovaries. Finally, fertilization is engineered in the Petrie dish. These procedures reveal the woman's reproductive body as passive, fungible material, a biological system that is broken up into its component parts[9] of uterus, tubes, eggs, endometrium, and hormone cycles that are worked upon by the *technités*. As Kathryn Russell notes,

> Individual women are to a great extent interchangeable; within certain medical parameters, one uterus or egg is as useful as another...We see that reproductive engineers are engaging in productive consumptions, expending their labor power in uniting the raw material (egg and sperm) with a functioning uterus to create a product.[10]

Human biological life is challenged-forth by the doctors either by placing the egg and the sperm in a Petrie dish or more aggressively by injecting the sperm directly into the egg (intracytoplasmic sperm injection). Thus, prior to being transferred to the womb of the mother or to the womb of the gestational surrogate, the embryo can be seen as being externally set into motion. All along the moving principle appears to be in the fertility doctor who tries to extrinsically energize a recalcitrant material that withholds its movement and resists the actualization of its potential. So, IVF does not merely give nature a push, as if taking a technological detour to safeguard something that was already poised to happen all along. Rather, here *techné* coerces nature to generate an embryo, attempting to trick the womb with hormone therapies into performing gestation and then medically supervising the growth of the fetus all along the way.

Most of the time implantation and thus gestation fail. For reasons that remain largely unknown, this part of IVF resists mastery.[11] When the process is successful, however, it follows the "invariable sequence" of pregnancy and birth whereby nature moves itself to completion, albeit with continuous medical assistance and monitoring. In view of so much medical assistance and technological intervention, we must indeed strain to see nature as the "source of being moved...in virtue of itself" (Aristotle 1941, 192b22–23), and as a "self-unfolding emergence" that is also a "going back into itself" (Heidegger 1998, 195)—and to overlook the role of the fertility doctor as the true *arché* of this movement.

The fertility doctor goes a long way to help ignite the spark of life. Even if the doctor's *techné* becomes more intrusive, however, selecting for particular genetic features such as sex, eye color, or "turning off" defective genes—and thus in a very limited way manipulating the form—the IVF embryo can be made serviceable only within enduring limits or *teloi*. As brought out by Heidegger's interpretation of Aristotle's *Physics*, "The limit is always what...defines, gives footing and stability, that by which something begins and is" (Heidegger 1977, 206, 1977, 8). Thus, the limit endures in the development of the embryo into a human being, growing arms and legs instead of wings or tails, for instance. Nonetheless, IVF shows that even as the doctor's medical incursions are limited by the specific set of possibilities involved in the reproductive matter itself, the doctor's push against those limits hints at the possibility that the ideal version of IVF in the future will erase all limits, enabling a technological production of the human being increasingly from the ground up.

Today medical incursions into human conceptions come closer than they ever have to cause life, to make it happen.[12] According to one male fertility doctor, "we are clearly helping those people to conceive that *mother-nature would have never allowed to conceive*" (my emphasis).[13] This view presupposes Aristotle's patriarchal ontology of sexual reproduction and also moves beyond it. In *The Generation of Animals*, Aristotle describes a relationship between an active male element and a passive female element (Aristotle 1941, 729a29–32), which can be transposed to today's doctor-patient relationship in the implementation of IVF. In accordance with Aristotle's views, in the matter of medicalized human conception, even today

> *the female, as female, is passive, and the male, as male, is active, and the principle of the movement comes from him.* Therefore, if we take the highest genera under which they each fall, the one being active and motive and the other passive and moved, that one thing which is produced comes from them only in the sense in which a bed comes into being from the carpenter and the wood. (729b11–19).

Aristotle's analogy intends to illuminate that just as a carpenter's *techné* (know-how) gives a bed its recognizable form, so it is a man's seed/semen that gives the human offspring its humanness. In the context of today's fertility treatments, however, this binary ontology that privileges male power over female power and the active principle over the passive principle receives an unexpected twist. The central role of the "active" semen is replaced by the technical tools of the fertility doctor who uses them on women's and men's reproductive nature alike as if this were passive matter. The fertility doctor's use of reproductive technologies places him in the unique world-historical position of coming as close as possible to artificially controlling the genesis of life, making human life happen. This tremendous sense of agency is implicitly underscored when fertility doctors, instead of nature or chance, are held responsible for IVF children who are born with some type of congenital impairment.[14]

As efficient causality becomes dominant in early modernity, so does the growing criticism of Aristotelian teleology. This dominance undermines the relevance of the final and formal dyad for setting limits and circumscribing change in natural and artificial beings. Today artificial (re)production takes its measure from the powers of the efficient cause, the doctor. Her influence, as the efficient cause of healing, can hardly be overestimated. According to Heidegger's interpretation of nature, in modernity "the *causa efficiens*... sets the standard for all causality" (Heidegger 1977, 7). In fact, the subordination of all three causes to the efficient cause becomes one of the defining features of the technological age (*Ge-stell*). As the technical manipulation of and experimentation with human reproduction continues to expand, Heidegger's theory of enframing can help us see where control over the spark of life lies—whether with nature or with medical technology.

Nature Without Limits

In 1939, almost 40 years before the first IVF baby was born, Heidegger seems to allude to the scope and impact of this reproductive technology. In "On the Being and Concept of φύσις in Aristotle's *Physics* B, 1," he writes:

> *Techné* can merely cooperate with *phusis*, can more or less expedite the cure; but as *techné* it can never replace *phusis* and in its stead become itself the *arché* of *health* as such. This could happen only *if life as such were to become a "technically" producible artifact*. However, at that very moment there would also no longer be such a thing as health, any more than there would be birth and death. Sometimes it seems as if modern humanity is rushing headlong toward this goal of *producing itself technologically*. (Heidegger 1977, 197; my emphasis)

His account of nature in "The Question Concerning Technology," more than 10 years later, seems to confirm this headlong rushing. He states that "the revealing that rules in modern technology is a *challenging-forth*..., which puts to nature the unreasonable demand that it supply energy that can be extracted and stored as such" (Heidegger 1977, 14). Moreover, as stored and storable, reproductive nature no longer "stands over and against" technology as an object that is occasionally exploited. In the age of technology "nature belongs, in advance, to the standing-reserve of orderability within enframing."[15]

We can see that Aristotle's view of nature as growth, a self-emerging and blossoming forth guided and drawn out by a given end, is supplanted by a "disclosive looking"[16] defined by production. In the age of efficient technological production there is no longer a stable *eidos* available anywhere, whether in the maker's mind, in social practices, or pregiven in the naturally enformed matter. In fact, technology's prowess is measured by its ability to efficiently overcome all naturally and culturally imposed limitations in an ongoing process of self-optimization. Andrew Feenberg sums it up as follows:

> Enframing... dissolves all traditions, the linguistic heritage, the fixed meanings on the basis of which people have engaged with the world in the past... Being becomes the object of pure will, the meaning of things derives from their place in the technological system rather than from a determinate *eidos*... Thus, our technological culture understands the world on the model of raw materials invested by ever shifting subjective plans. (Feenberg 2005, 21–24)

Despite the revolutionary changes that IVF has introduced in the conception and gestation of human beings, in the origin of human life itself, the dominant media and medical discourses (mis)represent IVF along Aristotelian lines as a process that merely gives nature a push, simply "completing what nature could not finish." This view conceals the revolutionary power of the practice.[17] The occlusion is revealed in the nostalgic and unwittingly Aristotelian cultural discourse about IVF that tries to bring it much closer than it actually is to vaguely familiar and ancient concepts of growth and making. Yet a further look into the actual phenomenon reveals an imposition upon nature, a challenging-forth that denatures it, "switching" the potential energy about, as Heidegger writes—in this case reproductive energy—trying to paradoxically delete and hold on to it.

As discussed in Chapter 3, IVF facilitates a host of reproductive techniques whose goal is to engineer human life rather than merely to nurture it into being. For instance, IVF is the gateway technology for the production of babies from "unborn mothers," that is, from aborted female fetuses whose collected ovarian tissue (matter) is harnessed into ova that could be stored or fertilized in a Petrie dish and eventually gestated into a child whose biological mother was never a person. On the other hand, cytoplasmic transfer is a new reproductive technology that "refreshes" an older woman's egg by combining it with the cytoplasm of a younger woman's egg in order to facilitate pregnancy for the older woman.[18] This procedure has produced babies with two living (genetic) mothers. Together, these two reproductive techniques reveal a real departure from sexual and low-tech reproduction because, as already argued, they help to disclose a world of fungible motherhood, a motherless world.

The Aristotelian causal framework no longer applies to these reproductive contexts because here all causality has been reduced to the fertility doctor as a source of motion, to the doctor's technical skill and medical apparatuses in making new life happen—increasingly without the limit of a specific form or *telos*. The primacy of this efficient cause is incoherent on Aristotle's model, according to which the formal and final causes necessarily define a thing's being. So, Aristotle's causal explanation of nature and technology cannot help us to make sense of invasive IVF procedures that presuppose the fungibility of people and things. Heidegger's account of technological revealing, however, casts nature as a heap of fungible raw material that lacks a being of its own and so borrows it, so to speak, from the production process.

The technical significance of producing children through IVF comes more clearly into view if we begin to see IVF not in isolation but rather *in continuity* with the drive for and development of other reproductive technologies, some of which I mentioned above and others already implicit in current practices and devices. Seeing the *continuity* across these reproductive procedures can open the door to seeing that the production of children through IVF signals a deeper articulation of technological enframing, one wherein the challenging-forth directed outward is also a self-objectification that reaches into the depths of human biology and culture. The increased normativity of this self-objectification shows up in the naturalization of reproductive technology, in the stubborn elision of the boundary that separates it from nature. The elision shows up, for instance, in the embryonic selection of secondary characteristics and in advanced surgical incursions into embryos for the purpose of producing healthier and more desirable babies.[19] Coupled with the gestation of babies in a gestational surrogate and soon, perhaps, in a manmade womb such as the one developed in Japan two decades ago,[20] these technologies push toward the elision of reproductive nature with the final goal of perfecting and personalizing offspring. The guiding preference is genetic ownership of the child.

Despite the growing popularity of IVF, the procedure still fails to produce a live baby in the majority of cases. When the procedure succeeds, the doctors receive some praise but the real praise goes to the technology itself. The public and media seem truly awed by the *technology* that was seemingly able to ignite the spark of life. They tend to attribute this spark to the power of technology and to having the right kind of technology more than to the *techné*, skill, or know-how of the fertility doctor. Yet despite the accolades for and reliance on reproductive technology, when it comes to the status of children produced through IVF, we refuse to (fully) acknowledge their technological heritage. We attribute the final result to nature. In the end, the significance of reproductive technology is undermined (or covered up) in real life as it is on the silver screen when sexual passion all too often takes the credit. For example, Kelly Oliver's feminist analysis of sexist and heteronormative Hollywood romantic comedies underscores this point when she writes, "these films reassure us that babies are not products of technology but rather of passion, and that sexual desire is necessary to fulfill the desire for babies" (Oliver 2010, 771). In real life, the attempt to cover up a child's origin in reproductive technology is routinely practiced in attempts to match the phenotypes of egg donors with those of the social parents so as to make the children seem "natural," that is, genetically linked to the social parents. It is also practiced in the refusal to tell children that they were conceived through IVF using donor eggs.[21]

What is at stake in this refusal is the preservation of biological ownership and its privilege over the nurturing aspect of parenting. The popularity of IVF reveals a cultural commitment to biological ownership, a desire to see ourselves somehow preserved in our offspring. Whatever the cultural connection may be, a prereflective and patriarchal "biological imperative"[22] calls forth the production of an offspring that *looks* like one or both of the parents, and this imperative demands the *naturalization* of IVF. In other words, the imperative demands the elision of what could be perceived as a natural deficiency in the would-be parents. Infertility demands that the extent and significance of the technological means necessary to correct the deficiency be minimized, even suppressed. So, what emerges is a *paradox* where the desire for nature feeds the desire for technology to seamlessly make us "natural."

As discussed in the previous chapter, this hankering after nature, or "mother nature," as a longing for an origin and a certainty that never were present—a "maternal fantasy"—can be seen as what Kelly Oliver calls the *mother-effect*—a concept inspired by Derrida's deconstructive genealogy according to which "the connection between mother and nature [is] a fable that covers over both technology and chance that are always already at the origin of humanity" (Oliver 2013, 55). This psychoanalytically driven concept describes "the result of the absence of the mother" (Oliver 2013, 57), an absence that is covered over. IVF participants (the women, doctors, and media) can be seen to reproduce the *mother-effect*, caught up in a play of affirming the significance of technology for conception and gestation, yet undermining this significance in the final product, calling it (mother) nature. In my view, the popularity of IVF brings out the *mother-effect*, just as the latter helps to cast IVF as increasingly normative. In fact, the *mother-effect* helps to make sense of what is going on today in the age of the reproductive enframing, with the "real mother" becoming increasingly absent as, likewise, her absence is covered over. The "real mother" (or, as I name her in Chapter 3, the *whole* mother) as the biological, genetic, and/or social point of origin for the child tends to vanish as she is replaced more and more, especially in conception and gestation, by IVF-based technologies of life.

We ought to radically rethink what we actually produce when we reproduce ourselves artificially. We fall into complacency as long as we continue to implicitly think of IVF along Aristotelian lines as merely giving nature a push and completing what nature is unable to finish. Heidegger's account of enframing is helpful in seeing IVF on a continuum with other reproductive technologies as ways to "challenge forth" human life and to control its production, more and more from the ground up, increasingly unrestricted by chance and by the limits of matter and purpose, imposed traditionally on "natural" and "technical" things. As long as we think of IVF technologies as giving us a piece of nature, the truly technological and revolutionary dimensions of the machinery and of the cultural outlook remain submerged and invisible.

Notes

1. There are other notable passages in Aristotle that resist the dominant interpretation that equates nature with form due to the view that potentiality ultimately resolves itself into actuality. For instance, see Aristotle's *De Anima*, book 2, where he underscores the unity of natural entities, the *ontological coupling* between matter and form. He writes, "Hence, too we should not ask whether the soul and body are one, any more than whether the wax and the seal are one, or universally whether the matter of each thing and that of which it is the matter are one" (412b6–9).
2. I am sympathetic to Carol Bigwood's critical reading of Aristotle's putative ontological misogyny, whereby the male semen in particular and the active, form giving, male principle, more broadly can be seen to appropriate (primary) matter, to potentialize it and so to delete it altogether. Potential matter is masculinized and as a result the woman's presence, traditionally associated with matter, is deleted. She becomes a fungible

resource. However, the disappearance of matter, as posited by Bigwood, is not consistently borne out by key passages in Aristotle's *Physics*. Aristotle seems to equivocate between matter as "relative" and "special" (194b10) and the more general description of matter as "the primary substratum of each thing, from which it comes to be without qualification, and which persists in the result" or "the immediate constituent... that taken by itself is without arrangement" (193a10). Thus, pre-enformed matter seems to endure and to be significant, albeit not as significant as its corresponding form.

3. I am indebted to Professor J. J. Glanville's graduate course on Aristotle (San Francisco State University, 1994) for the clarification of this important point. See also Trish Glazebrook, "From φύσις to Nature, τέχνη to Technology: Heidegger on Aristotle, Galileo and Newton," in *The Southern Journal of Philosophy* 38 (2000): 95–118. In the same context Glazebrook adds that "A thoughtful account of Aristotle looks not to prime mater to think the enigma of generation" (103). This can be seen as a criticism of Carol Bigwood's account of prime matter as the maternal element that is suppressed in Aristotle's account of generation (see Bigwood, *Earth Muse*, ch. 5).

4. While *techné* is said to imitate nature, Aristotle sometimes tries to makes sense of nature by looking at *techné*. See book 2 of Aristotle's *Physics*, especially ch. 8, 199b 25–30; Aristotle, *Physics*, trans. R. P. Hardie and R. K. Gaye, in *The Basic Works of Aristotle*, ed. Richard McKeon (New York: Random House, 1941).

5. Schummer's basic point seems to be that humans, for instance, learned to become effective hunters and build effective hunting tools by watching animals hunt other animals.

6. On the relationship between *telos* and *peras*, see Heidegger, "On the Essence and Concept of φύσις *in Aristotle's Physics, B 1*" (192, 206).

7. Andrew Feenberg, *Heidegger and Marcuse on the Catastrophe and Redemption of Modernity* (New York: Routledge, 2005), 7.

8. For a detailed account of this harnessing, see the first section of Chapter 3, above.

9. For more details on the decontextualization, reduction and fragmentization in technical processes, see Andrew Feenberg, "Impure Reason," in *Questioning Technology* (New York: Routledge, 1999).

10. Quoted in Wendy Lynne Lee, "Reproductive Technology and the Global Exploitation of Women's Sexuality," in *Contemporary Feminist Theory and Activism: Six Global Issues* (Broadview Press, 2010), 64.

11. Karen Dawson and Peter Singer, "IVF Technology and the Argument from Potential," in *Embryo Experimentation* (New York: Cambridge University Press, 1990), 78.

12. See Joachim Schummer's account in "Aristotle on Technology and Nature," *Philosophia Naturalis* 38 (2001), 113. According to him "there is no ontological difference between these (artificial) products and natural products, because both in chemistry and in nature the dominant principles of generation, i.e., material cause and efficient cause, are the same (i.e., heat and cold)." Pace Schummer, the efficient causality in nontechnological conception and conception through IVF differs widely.
13. Dr. William Gibbons in the documentary film *My Future Baby: Breakthroughs in Modern Family* (2012), directed by Brigitte Mueller.
14. On the responsibility of the doctor see L. Rapaport, "IVF Linked to Birth Defects and Childhood Leukemia," in *The Huffington Post*, February 4, 2016. See also J. Kluger and A. Park, "Frontiers of Fertility" in *Time* magazine, May 30, 2013. It seems that IVF babies who are born of parents who could not reproduce themselves without technological intervention owe their being to their biological parents as much as they do to the fertility doctor. The two are co-archic.
15. Martin Heidegger, "Overcoming Metaphysics," in *The End of Philosophy*, ed. Joan Stambaugh (Chicago: University of Chicago Press, 1972), 104.
16. See Richard Rojcewicz, *The Gods and Technology: A Reading of Heidegger* (Albany: State University of New York Press, 2006), 68.
17. See Hans Jonas, "Technology and Responsibility: Reflections on the New Task of Ethics," *in Society, Ethics, and Technology*, 4th ed., ed. M. Winston and R. Edelbach (Wadsworth, 2012), 121–132.
18. Karey Harwood, *The Infertility Treadmill: Feminist Ethics, Personal Choice, and the Use of Reproductive Technologies* (University of North Carolina Press), 12.
19. Michael Sandel, "The Case Against Perfection" in *Society, Ethics, and Technology*, 4th ed., ed. M. Winston and R. Edelbach (Wadsworth, 2012), 329–340.
20. See www.nytimes.com/1996/09/29/the-artificial-womb-is-born.html.
21. See, for instance, G. Pennings, "The Right to Choose Your Donor: A Step Towards Commercialization or a Step Towards Empowering the Patient?" in *Human Reproduction* 15, no. 3 (2000): 508–514 The ongoing debate about whether or not it is best to tell children born of donor eggs the truth about their genetic origins indicates that secrecy is widely preferred. The literature on this debate is vast, including personal, medical, and academic accounts. See www.dreamababy.com and http://www.abc.net.au.
22. See Robyn Ferrell's *Copula: Sexual Technologies, Reproductive Powers* (Albany: State University of New York Press, 2006), 32.

CHAPTER 5

On the Harnessing of Birth in the Technological Age

Abstract This chapter applies the concept of the reproductive enframing to technophilic and technophobic approaches to childbirth. It focuses on the over medicalization of childbirth, especially as it takes place in the form of cesarean sections, and on nontechnological forms of childbirth such as Lamaze. With the help of feminist phenomenology the chapter shows that, although these two examples of technophilic and technophobic childbirth appear to be in opposition to each other, they in fact express different stages of the reproductive enframing.

Keywords Reproductive enframing · Technophilia · Technophobia · Childbirth · Lamaze

In this chapter I focus on reproductive technophilia and reproductive technophobia in childbirth. I interpret technological and nontechnological, but *techné*-centered, childbirth through a phenomenological lens. Nontechnological childbirth is not synonymous with natural childbirth insofar as the latter usually refers to an automatic, biological process that simply overtakes the woman and activates a putatively innate birthing knowledge. I eschew this essentialist and patriarchal interpretation because it collapses childbirth into a passive bodily function rather than viewing it as active work (Held 1989, 364). My working assumption is that all approaches to childbirth are always already culturally conditioned and involve some deliberation and choice on the part of the pregnant woman.

Despite important differences, technological and nontechnological childbirth both reflect contemporary western cultural norms of order(ability), control, and optimization. The technological model is technocratic because it "defines the body as a machine... and insists that, like a machine in a factory, the laboring body should produce its product within a specified amount of time; and if not, this birthing machine is obviously dysfunctional and in need of... repair" (Davis-Floyd 1998, 259). Its anti-technological counterpart, the Lamaze method, traditionally emphasizes acquiring birthing skills and minimizing technical interventions. Like its technical counterpart, however, it strives to conquer nature. Noting that both models reflect traditional masculinist commitments to aggression and control, I rethink these childbirth scripts in light of Heidegger's theory of technology. Applying Heidegger's concept of enframing (*Ge-stell*), I advance a feminist phenomenological interpretation that shows how these two childbirth models help to disclose the technological age—and are already caught up within it.

The World as Resource

In "The Question Concerning Technology" Heidegger uses the term *enframing* (*Ge-stell*) to capture what he considers to be the essence of the technical age. His account, which is ontological rather than sociological, claims that enframing is "nothing technological," but a historical (*geschichtlich*)[1] mode of revealing (*aletheuein*) unique to the West. It refers to an inherited cultural outlook dominant in advanced, industrialized countries according to which nature and people are reduced to raw materials, fungible units of consumption and commerce. Enframing naturalizes a value system geared toward continuous production, efficiency, and enhancement. Heidegger writes that, as a consequence of this reduction of everything to fungible resources in the age of enframing, people's lives increasingly lack rootedness and historical belonging, or what he calls maternal protection (*Hut*) (Heidegger 1949, 46).

According to this theory, things show up only insofar as they have the potential to be ordered, that is, "regulated and secured" (Heidegger 1977, 175). The phenomenology of enframing[2] describes a relationship between an attitude of imposition or "challenging-forth" and what this attitude discloses, the world as a mass of fungible raw materials, resources, or "standing-reserve" (*Bestand*) awaiting optimization. The dominant value embodied by the attitude of "challenging-forth" is a constant "driving on to the maximum yield at the minimum expense... Everywhere, everything is ordered to stand by, to be immediately at hand, indeed to stand there just so that it may be on call for a further ordering" (Heidegger 1977, 15). Enframing "orders what presences as available and reportable... as standing-reserve" (Heidegger 1949, 40). People and nature are reduced, each in their own way, to units of energy and information that are then "unlocked, transformed, stored, distributed and switched about" (Heidegger 1977, 16). What doesn't show up as a resource or a potential resource, what cannot be challenged-forth, simply does not show up at all.

According to Heidegger's famous phenomenological analysis of the power plant on the Rhine (Heidegger 1977, 16), the river no longer gathers local traditions and lore, but is primarily seen as *potential* energy in the form of a water resource for the power plant. It is built into the dam, rather than the dam being built into the river. Although multiple technologies are used in the enframing of the power plant on the Rhine, challenging-forth is not restricted to technological contexts, as we shall see shortly.[3] As explained in previous chapters, Heidegger's phenomenological approach peels back the layers of the mainstream instrumental understanding of technology that sees technology as a neutral tool. This instrumental attitude is "limited to a subjective set of conditions" (Heidegger 1977, 4–6) that it presupposes but cannot explain. One such significant subjective condition is the still dominant seventeenth-century Cartesian and mechanistic assumption that individual rational subjects control inert objects, that is, those entities perceived to lack reason and agency. The mechanistic assumption that reason and "motion [were] external to matter"[4] promoted a widespread instrumental attitude that is still popular today.[5]

Since instrumentalism is stuck in modern subjectivism it is unable to recognize this historical bias. For example, it cannot tell us *why* rationality, objectivity, and maximizing output matter, but simply tells us *that* they matter. It cannot explain why today norms such as efficiency, order(ability), and control are dominant and not others—such as tranquility, piety, or spontaneity.[6] The instrumental subject fails to notice that what she does not master is the "will to mastery," and that in multiple contexts she now treats herself as an object. Iain Thomson puts it as follows:

> Once modern subjects dominating an objective world begin treating *themselves* as objects, the subject/object distinction itself is undermined,

and the subject is thereby put on the path toward becoming just another resource to be *optimized*, that is secured and ordered *for the sake of flexible use*. (Thomson 2005, 60)

This result is mirrored in how entities are disclosed. Heidegger writes:

> Thus when man, investigating, observing, ensnares nature as an area of his own conceiving, he has already been claimed by a way of revealing that challenges him to approach nature as an object of research, until even the object disappears into the objectlessness of standing-reserve. (Heidegger 1977, 19)

Yet, another way is possible:

> The field... appears differently than it did when to set in order still meant to take care of and to maintain [*hegen und pflegen*]. The work of the peasant does not challenge the soil of the field. In the sowing of the grain it places the seed in the keeping of the forces of growth and watches over its increase. (Heidegger 1977, 14–15)

This nonenframed approach points to a "technique that works with nature and the material, allowing beings to come forth at any given time, while modern technique strives to order, confine, control, and then to 'challenge' nature to produce" (Klawiter 1990, 71). In "Science and Reflection," Heidegger clarifies this phenomenological difference in our understanding of nature when he writes that nature as resource "is only one way in which nature exhibits itself" (Heidegger 1977, 116). Human beings order and "drive" nature and technology forward but never produce how nature (or being) reveals itself. We always presuppose a specific understanding of nature, and we do this in our practices and not just in our heads. Blindness to this presupposition is a symptom of the modern subjectivist bias, a lack of awareness that we are historically determined to relate to nature in a controlling and mechanistic way.[7] This unacknowledged limit of the instrumental position covers up as it also points to the enframing.

According to Heidegger the instrumental understanding gets stuck in the technologies themselves and is therefore phenomenologically naïve, providing only a "surface definition of technology" (Ihde 2010, 31). Two examples of such naïve phenomenological views are *technophilia*, the unqualified love of technology, and *technophobia*, the fear and hatred of all things technological. Although seemingly different, these views are fundamentally similar in that both see technology along instrumentalist lines as merely a neutral object pushed around by a rational subject. Heidegger critiques these views as reactionary and avoids discussing the pleasures and terrors of technology that they, respectively, endorse. He does however state that *technophilia* and *technophobia* perpetuate the "stultified compulsion to push on blindly with technology or, what comes to the same thing, to rebel helplessly against it and curse it as the work of the devil" (Heidegger 1977, 25). In the arena of reproduction, these views are embodied in the technocratic and anti-technological models of childbirth.

Technophilic Childbirth and the Call for Pain Relief

In the next two sections I situate Heidegger's phenomenology, as described in the section above, in a feminist context. In my feminist phenomenology I show how the use of modern childbirth tools disclose women's reproductive bodies first as *objects* and later as *resources*. While staying close to Heidegger's work I also maintain a critical distance from it. For instance, I agree with Lorraine Markotic, in "Paternity, Enframing, and a New Revealing," that in view of the later Heidegger's recurring discussion of *poiésis* as a bringing-forth and of *physis* as "the highest form of *poiésis*" (Markotic 2016, 130), his silence on birth is surprising and troubling.[8] I return to this silence in the next chapter.

In a sense, philosophers such as Hannah Arendt can be seen to have taken up the themes of bringing-forth and throwness or *Geworfenheit*, the latter as natality, a political coming to be, the emergence of a singular point of view (Arendt 1958, 178). However, Arendt's philosophical work and that of other philosophers on natality and sexual difference[9] inspired by Heidegger differ from my own insofar as none provides a detailed and sustained phenomenology of the use of reproductive technology in childbirth.[10] In view of the overmedicalization of birth in the last century, and especially in the last few decades, this work seems urgent. Let me emphasize that it is undeniable that some advances in modern reproductive technology have made childbirth much safer for scores of women and newborns than it was in previous centuries. However, these advances have also promoted a social infatuation and overdependence on reproductive technologies, especially drugs, that often interfere with the well-being of mother and child. The increased use of advanced reproductive technologies (ARTs) continues to shift attention away from the autonomy of the mother and onto the newborn, whose safe arrival into the world is the primary medical objective. Realizing this objective requires the use of multiple technological interventions.

According to feminist Suzanne Arms, "the predominant technology in childbirth from the beginning of the 1850s to today has been the use of drugs" (Arms 1994, 74). Today the high rates of cesarean births in the U.S. (Michaels 2014, 127), including elective cesareans,[11] mark the triumph of drug technologies, the marginalization of a pregnant woman's birthing *techné* or "know-how." My feminist phenomenological interpretation of the history of that loss notes the inversely proportional correlation between the medical (re)ordering *of* birth and women's marginalization *from* birth. Between 1850 and the late 1960s, new medical technologies in the U.S. and Europe changed birth from a woman-centered process to a technology-centered process focused on producing tools, developing experimental techniques, and placing them in the hands of male obstetricians and their medical staff. Traditional childbirth with a midwife was on its way out, and along with it a woman-centered childbirth was being marginalized; that is, the experienced hands of (mostly female) midwives working with the laboring woman to bring a baby forth were displaced in favor of male-driven theoretical medical knowledge. This knowledge enjoyed high social status but lacked experience, relied on poor theoretical instruction, and used devices to challenge-forth the birth (Louden 2002, 297). The love of obstetric tools, moreover, coupled with the perception that most pregnant women are "unable" to birth naturally (i.e., without technical intervention) had, by the beginning of the twentieth century, framed birth in the U.S. as a pathological event.[12] The remedy for this pathology was believed to hinge on using more medical technology. This view garnered support from, even as it reinforced, a growing technophilic attitude. After multiple medical interventions, childbirth became more technologized but no "cure" was found. What was found, however, were multiple ways to enhance[13] or optimize it, that is, to switch it about, regulate it, secure it, and expedite it (Heidegger 1977, 16) for purposes that often served neither mother nor child, and transformed birth into an "obstetrical manufacturing process" (Davis-Floyd 1998, 169).

Modern childbirth is believed to have been initiated in the 1730s when a Scottish apothecary named William Smellie perfected the obstetric forceps (Rich 1986, 141–142). For decades the forceps were used sparingly and successfully, often saving both the baby's life and that of the mother in emergency situations. By the end of the eighteenth and beginning of the nineteenth century, however, forceps were used for a variety of reasons: to ease the tedium of delivery, to save the surgeon's time, and simply because surgeons were infatuated with a new technology that symbolized progress. This overuse often resulted in botched births, damaged mothers, and dead babies (Wajcman 1991, 65). Nonetheless, the technology was widely available and could not be put back in the box; in this respect it was out of control.

In *Recreating Motherhood*, Barbara Katz Rothman writes, "When childbirth became a medical event, women lost control over their own birth experiences. The medicalization began with the eradication of midwifery as a profession, and continued largely unabated until the homebirth and midwifery movements of the 1970s" (Rothman 2000, 154). The poet Adrienne Rich, in *Of Woman Born*, underscores this view: "No one disputes that within recorded history, until the eighteenth century, childbirth was overwhelmingly the province of women" (Rich 1986, 131). Although midwives controlled the birth process, this control did not challenge the hegemony of patriarchy, whose stringent imposition of motherhood on all women and control over the fate of their offspring continued unabated. Even the midwives' control over childbirth was not to last. By mid-nineteenth century most women in the U.S. were no longer perceived as being "capable of standing the pain and stress of labor," thus if "women were no longer able to withstand the pain of labor, new methods...were needed" (Loudon 2002, 343). By the beginning of the twentieth century, three interconnected features defined modern childbirth, and continue to do so today: "new methods of pain relief, massive intervention in normal labors, and admission to hospitals for as many cases as possible" (Loudon 2002, 343).

The beginning of a biomedically constructed birth can be traced back to the marginalization of the midwife in Europe and the simultaneous rise of forceps overuse. This overuse was initially (and remained primarily) connected to the sheer love and control of technological innovation by male doctors, and later to the doctors' desire to "expedite" (Heidegger 1977, 15) birth. According to multiple sources, including the disparate works of Loudon, Rich, and Wajcman, this overuse was not *primarily* motivated by a concern with relieving women's pain in labor. A lack of concern with women's well-being, especially poor and working class women, is not at all surprising considering the deeply patriarchal and classist norms at the time. In fact, monopoly of the forceps was tightly guarded during the seventeenth century by its male inventors—the Chamberlain father-and-son team—who used it selectively and for a high price. Midwives, who were almost exclusively female, were prohibited from using technology during childbirth, including forceps. This led to their professional demise.

According to my Heideggerian interpretation, the overuse of forceps begins to frame the birthing woman as "an object of research" (Heidegger 1977, 19) who *primarily* served the technology rather than the technology serving her. Nor did it seem to serve the interests of the obstetrician qua obstetrician because he was (for the most part) not alleviating pain but causing it, often along with severe injury to both mother and child.[14] Moreover, the high failure rate of the technology did not deter its use, which remained controversial through the nineteenth and twentieth centuries. This is underscored by J. Drife's account:

> The obstetric forceps have remained controversial throughout their history, and in the 20th century a major reason was that they were used too readily and sometimes without the necessary skill. At the end of the 19th century and during the first decades of the 20th, obstetrics formed a major part of general practice, and *in the interests of efficiency a busy general practitioner would often apply the forceps rather than waiting for a normal delivery.* In response to a plea for conservatism in the *British Medical Journal* of 1906, several general practitioners wrote attacking elaborate aseptic precautions as unnecessary and normal delivery as impossible for "civilised" women. This epidemic of unnecessary intervention was one of the reasons why the maternal mortality rate in Britain in 1935 was the same as it had been at the beginning of Queen Victoria's reign. (my emphasis)[15]

Today the use of forceps has sharply declined, being widely replaced by vacuum extractions. However, in the nineteenth century the use of forceps was popular and often coupled with the use of ether and chloroform. By the early nineteenth century two male physicians, Clement and Leake, definitively "establish[ed] the lithotomy (lying-down, therefore passive) position as the preferred one for women in labor" (Rich 1986, 146). Then, in 1847 Sir James Young Simpson discovered the anesthetic properties of chloroform (Arms 1994, 55). This technology consolidated the (male) physician's control of birth by rendering the woman useless and unconscious, a mere medical resource. The drug erased her pain and memory of birth. Adrienne Rich describes the experience in the following words: "Floating into euphoria as she reclined on her back on the doctor's table, a woman yielded everything—her attention, her effort, her responsibility, and her sense of protection toward her baby—to the authority of the physician" (Rich 1986, 56). Rich then sums up the patriarchal demeanor of this procedure as follows:

> At the onset of labor, the woman was placed in the lithotomic (supine) position, chloroformed, and turned into the completely passive body on which the obstetrician could perform as on a mannequin. The labor room became an operating theater, and childbirth a medical drama with the physician as its hero. (Rich 1986, 170)

But more is going on. Rich's eloquent account assumes a Cartesian subject/object framework in which the physician is cast in the role of a rational yet unfeeling doctor and the woman is relegated to the object position. In view of Heideggerian enframing, however, Rich's quote allows us to glimpse the emergence of a new hero. This new hero is the drug itself, whose deployment can be seen as an early reframing of both physician and woman as resources. If we view chloroform on a spectrum with its significant successors, twilight sleep and the epidural, we notice how the use of chloroform begins to shift the disclosure of the woman's body as *more fungible* than did the limited use of forceps, something her body could oppose or reject.

Using Heidegger's account of enframing, as described in the previous section, allows us to see the difference between the body as *object* and the body as *resource*. One of the essential features of the technical age, according to Heidegger, is that entities are no longer, as in modernity, disclosed as objects or things that stand over and against subjects (Heidegger 1977, 17) defined by relatively fixed boundaries, (social) uses, and purposes. A *resource* is an object that "disappear[s] into the objectlessness of standing-reserve" (Heidegger 1977, 18) and is pliant, expeditious, that is, it "drives on to the maximum yield at the minimum expense," open to multiple purposes, serves a (technical) network that drives toward continual enhancement. The use of chloroform reveals many of these features. It enhances the obstetrician's access to women's reproductive bodies in a qualitatively deeper way than did the forceps before it. While it removed the pain of childbirth, thereby satisfying the desires of many women at the time, it also rendered them unconscious, thereby opening more experimental opportunities for the physicians. The withdrawal of the woman's moving, affective body enabled physicians to experiment with the cesarean technique. In fact, chloroform contributed to the expansion of modern cesarean surgery,[16] which in turn led to a mass preference for physician-controlled birth in a hospital setting. One significant limit remained to be reckoned with, however: the body's high sensitivity to death from overdosing. Due to high numbers of accidental mother and infant deaths, the use of chloroform in childbirth was eventually replaced with twilight sleep, and was finally discontinued.

The use of twilight sleep framed the pregnant woman as *object*. By 1914, before terrible side effects surfaced, twilight sleep induced by morphine and scopolamine became American women's preferred drug for both cesarean surgeries and nonsurgical hospital births. It offered the same benefits as chloroform and did not accidentally kill as many women. Due in part to a lack of proper medical training in administration and monitoring, however, it induced hallucinations and violent outbursts in many women. Women's wrists, elbows, and knees were cuffed to their hospital beds to prevent self-injury and injury to the fetus and staff (Arms 1994, 78). In order to conceal the cuff scars from their husbands—who were not allowed in the hospital rooms—the hospital staff switched to lambskin straps. Even with use of this softer material some women wore away much of the skin around their wrists from thrashing (Michaels 2014, 15). Many women were routinely left alone for hours, tied to their hospital beds.

During a nonsurgical twilight birth the nurse is routinely asked to press, even lean, on the mother's belly in an effort to push the baby out while the physician pulls the baby from the birth canal with forceps. Twilight sleep, like chloroform before it, ensured that the woman was not a conscious partner in birthing her child/ren, whatever the method of delivery. According to Sunnye Strickland, a pioneer in natural childbirth, many obstetricians "had not seen women awake for a birth except as a mistake, either a precipitous delivery or their own late arrival at the hospital" (Zwelling 2001, 17–21). This was reported to be the case as late as the late 1960s. Women's mechanized labor had become the norm. Two problematic aspects of twilight sleep were the mother's birth amnesia and the drug's adverse effects on the newborn's central nervous system. Many women felt bewildered and upset when presented with a newborn whom they had no memory of birthing. Some women reported thinking that they were still pregnant. Many of the newborns were drowsy and some needed resuscitation. Despite the adverse side effects, twilight sleep was used in the U.S. until the early 1970s.

This brief hermeneutical and phenomenological overview suggests that twilight sleep did not ultimately serve the woman or the fetus; rather, both of them served the drug and its relationship to a hospital network that orders physicians and nurses to expedite birth. The drug was supposed to serve multiple ends, such as removing pain, controlling the birthing process, and rendering the body a docile resource for obstetric intervention. Docility is not achieved, however; often and unpredictably the woman struck back, as her physical and psychological reactions were messy and obtrusive, reinserting her agency and affect, however poorly, into the process. These responses show her as *object* but not as *resource*. We shall see shortly that the use of the epidural demotes the position of the obstetrician as a subject set over and against the woman as object, and succeeds where chloroform and twilight sleep failed. It renders the woman quiet, conscious, and docile.

In the mid-1960s, the electronic fetal monitor (EFM) and the first version of the spinal epidural were widely introduced in hospitals throughout the U.S. The EFM band encircles the woman's belly and restricts her movement, while the EFM machine prints out a strip with markings that represent the baby's heartbeat. Despite the fact that "numerous studies have shown that continuous EFM monitoring does not improve birth outcome," pregnant women continue to believe that the EFM monitor somehow protects the fetus (Cartwright 1998, 244, 250, 248). The actual benefits of EFM are to protect doctors in court, promote hallway obstetric conversation, and allow for remote physician monitoring (Cartwright 1998, 244–245).

When coupled with EFM, the epidural foregrounds the resource status of the woman's reproductive body. The epidural is "an anesthetic injected into a space inside the spine, blocking nerve transmission. Epidurals were first touted by doctors and nurses because they made for quieter labor wards than did twilight sleep" (Arms 1994, 80). In addition to alleviating pain and inducing a state of euphoria, it allowed mothers to remain awake during surgery[17] and to immediately hold their babies afterward. Epidurals reduced pain and the number and intensity of contractions. Intense contractions were usually brought on by the drug Pitocin, administered to catalyze labor or to activate a stalled labor. The drug is administered through a heparin lock, a collection of small tubes attached to a catheter and inserted into the arm. The heparin lock is a preemptive measure mandatory in most hospital births, a central part of the technical network that promotes a biomedically constructed technocratic birth. Epidurals slowed down labor, diminished feeling below the ribs, and compromised a woman's ability to push her baby out without more technological interventions, such as episiotomies, vacuum extractions, or cesarean surgery. The recent increase in the use of epidurals (Michaels 2014, 11, 144) seems to have contributed to an increased rate in cesarean sections,[18] especially elective cesareans, the most efficient delivery option today.

The epidural reveals the woman's reproductive body as a *resource*. It renders this body fungible while itself withdrawing, remaining concealed. It can be administered continuously and in varying amounts to suit individual preferences. It is flexible, efficient, and carries little to no risk of death. Because the woman remains awake it gives the impression of birthing "normalcy" by comfortably mimicking natural birth, thereby enhancing "docility," or the body's capacity to be "subjected, used, transformed and improved" (Foucault 1977, 136). By comparison, chloroform and twilight sleep appear intrusive and primitive. Thus, the more invisible and "normalizing" the technology, the more fungible the woman becomes. The body's resistance as an object—the unruly, embarrassing "corpse" of the still living woman—has now disappeared.

The epidural also reveals the woman as a resource because, as noted in the first section, it allows her to treat herself as object, but now she can do so safely and as a matter of preference; that is, it is her call that the anesthesiologist must answer. The disappearance of the obstetrician as subject is evidenced not merely by the woman's ability to demand the drug, but by the drug itself, the real source of power in this medical situation. While under the spell of the epidural, the laboring woman is aware that things are happening to her and she may even feel some pleasure, but she does not feel any pain or effort, all of which would animate the experience and give it personal traction. Many women feel "detached from the experience of birth" (Michaels 2014, 129), and this renders the experience anonymous. The anonymity can be seen as resulting from the drug-induced passivity that leads to fungible treatment. Viewed alone, in fact, the epidural goes far in making what I call reproductive enframing visible. Reproductive enframing[19] aims to capture the normalization of the medical interrogation of women's reproductive bodies, casting them as fungible in that they receive the same medical treatment and are having increasingly standardized, interchangeable birthing experiences. This fungibility can be seen when the epidural removes enough feeling below the torso that a cesarean section appears imminent—and becomes the next (seemingly) necessary step. In this sense, women's bodies and birthing experiences are treated as interchangeable.

Reproductive enframing in childbirth is most visible as the ensemble of hospital drugs and technologies working together. The technical "interlocking paths" formed by the hospital clock, the heparin lock, Pitocin, and the EFM presuppose the enforced technique of the lithotomic position and culminate in the use of epidural anesthesia. Together they transform childbirth into an "obstetrical manufacturing process" that is primarily served by the woman's reproductive body rather than serving it. Hospital birth is closely monitored. If "progress" does not fit into a narrow time-frame, usually 24 hours or less, physicians will pressure women to have a cesarean section, often citing signs of fetal distress. It is difficult for the laboring woman, especially if she has no medical experience, to confirm or disconfirm the presence of fetal distress. The citing of fetal distress sets her up in opposition to her fetus and creates a maternal-fetal conflict, a situation in which, according to the obstetrician, the interests of the mother are seen to oppose the interests of the fetus. In order to avoid this conflict, women who might otherwise have chosen a natural birth in the belief that it is more beneficial to the baby, and to their bonding with their newborns, often surrender their will to the will of the doctor and opt for surgical intervention.

A constant reliance on technical intervention would seem to promote the deskilling of the obstetrician's hands-on "know-how" necessary to help with the twists and turns of labor. This makes more technical intervention seem like an attractive option. On this picture, women and their doctors become part of a medical network that neither one of them controls. While the doctor can still be seen to animate the technical network, the doctor is in fact ordered around by webs of administrative protocol and insurance liabilities that also frame her as a resource. The erstwhile subject-object relationship no longer holds.

Radical and socialist feminists[20] point out that the reason behind women's choices of medicalized birth is the internalization of patriarchal norms that have not been successfully identified and/or overcome.[21] In other words, too many women fail to think critically about the patriarchal norms and values they identify with, thus reproducing a patriarchal synecdoche, one that objectifies[22] women by reducing them to their reproductive bodies and then exploiting these bodies. Barbara Katz Rothman's influential book, *Recreating Motherhood*, for instance, is predicated on exposing the twin domination of pregnancy and childbirth by patriarchal ideology and technological ideology. The former sees the child as a "product... a seed planted in woman" while, according to the latter, "*technology is* used to problematically frame mother and fetus as separate beings; nesting Russian dolls, one inside the other" (Rothman 2000, 160; my emphasis). While new reproductive technologies undoubtedly contribute to more widespread and uncritical perception of the mother and the fetus as separate beings, Dorothy Roberts remarks that this separation was already in place during the years of slavery in the U.S. She recounts that slave owners used to whip pregnant black female slaves by asking them to lie face down in a hole that was dug just big enough to fit their pregnant bellies (Roberts 1999, 40). The fetus, which was perceived as a valuable product or *object* by the slave owner, was thought to be separated and protected from the harm inflicted on the mother. Roberts interprets this ferociously racist and misogynist practice as "the most powerful image of maternal-fetal conflict" (Roberts 1999, 41) that she discovered in all her research on reproductive rights and reproductive justice.

While I agree with Roberts that this practice illustrates a horrific treatment of the pregnant woman, I do not agree that the separation that it introduces *is* an example of the maternal-fetal conflict. In my view, it can be seen as setting a historical and cultural precedent for the maternal-fetal conflict invoked by obstetricians today. According to Roberts the maternal-fetal conflict describes "the way in which law, social policies, and medical practice sometimes treat a pregnant woman's interests in opposition to those of the fetus she is carrying" (Roberts 1999, 40). As mentioned earlier in this section, for example, a laboring woman in the U.S. who is informed by her obstetrician that her fetus is in distress, and who refuses her doctor's advice to have a cesarean delivery, is said to enter in an adversarial relationship with her fetus. This is how things look from the perspective of the obstetrician and the law. While the obstetrician's goal is to alleviate this conflict in order to benefit the fetus, his goal is not to injure the woman. When the slave owner whips the slave mother, however, his goal is to inflict harm on the woman. Thus, the conflict is between the owner (the subject) and the slave (the object). There is no initial conflict between the mother and the fetus; the fetus is merely incidental to this conflict. The owner seeks to protect the fetus in order to protect his economic interests, that is, "the production of future profits" (Roberts 1999, 41).

On my Heideggerian reading, the separation invoked by Roberts casts both mother and fetus as *objects* who are set over and against a controlling *subject*, the slave owner, who hurts them both in different ways insofar as the physical trauma experienced by the mother reverberates throughout her body, including the fetus, who is enmeshed with her body. The domination is somewhat akin to—though more gruesome than—the fright and dehumanization inflicted on pregnant women who, when administered twilight sleep, were shackled to their beds. The use of this patriarchal practice (shackling) and medical technology (the drug) together treat the fetus *as if* she were separate from the mother. Robert's interpretive insight allows us to see that in addition to patriarchal and technological ideologies, the (over)use of medical technology that today frames the mother and fetus as separate beings can also be seen to connect with racist ideology from the days of slavery.

Yet patriarchal and racist ideologies per se cannot tell us *why* the norms of order, efficiency, and control continue to matter; they simply tell us *that* they matter. Male and female obstetricians today are ordered to implement technological intervention by anonymous hospital policies, a network of insurance liabilities, and customer satisfaction that no one controls but that control everything. Thus, a normative desire for order (ability) and control is deployed in the medical network *as* the network itself that, as Heidegger writes, makes sure that the "regulating itself is, for its part, everywhere secured" (Heidegger 1977, 16).

Technophobic Childbirth and the Upside of Pain

Despite the increase in technological births today,[23] recent history in the U.S. shows a resistance to this technological trend. Starting in the 1960s and through the late 1970s, American women pushed back with the help of the Lamaze method. Lamaze taught them techniques for birthing without technology. This pushback made progress for a while, especially during the rise of social movements in the U.S. that included feminism. On my Heideggerian reading, however, this method reflects central features of enframing.

I suggest that women today could benefit from thinking about drug-induced technological birth as being fundamentally similar to the drug-free or low-tech alternative. Neither one achieves what Heidegger calls "a free relationship to technology" (Heidegger 1977, 4, 34). The traditional and technophobic Lamaze method reflects, like the technophilic method it criticizes, some of the essential norms of enframing: orderability and control. These two approaches can be seen to express different stages of reproductive enframing. The overmedicalized approach enframes by inserting the pregnant woman into the system as an *object* of technical action. The Lamaze approach enframes by making the pregnant woman the *subject* of technical action, a kind of laborer who is working to give birth. Since both approaches presuppose and promote order, control, and optimization, each in its own way projects the respective subject and object as *resources*, thereby undermining the distinction.

Grantley Dick-Reed's *Childbirth Without Fear* (1933) and Fernand Lamaze's *Painless Childbirth* (1956) popularized a nonmedical and somewhat technophobic approach to childbirth.[24] I am not suggesting that all or even most women in the U.S. subscribed to Lamaze during the 1960s and 1970s, but enough women did that it became a dominant and recognizable trend, especially among middle- and upper-class white women who presumably had the necessary time and money to pursue alternative birthing methods. In "The Political 'Nature' of Pregnancy and Childbirth," Candace Johnson shows that privileged women in North America—that is, white middle- and upper-class women (LaChance Adams and Lundquist 2013, 194–195)—make up the majority of women who pursue nontechnological and nonmedicalized births, with or without a midwife. According to Johnson, it seems that the challenge to "the orthodox view of medicalization...was launched (and sustained) largely in the interests of privileged women." This came as a response to "the problem of medicalization that seem[ed] to apply disproportionately to privileged women... (S)ome of the most serious pronouncements of medical interference in pregnancy and childbirth...come from women of considerable privilege and authority." (LaChance Adams and Lundquist 2013, 200). Their rejection of overmedicalization is an exercise of power, a rejection of external control and domination; it is not, however, a rejection of medical assistance, if needed.[25] As we shall see, this point applies nicely to the Lamaze woman. Johnson's feminist analysis of birth culminates in the interesting claim that privileged women who pursue nontechnological birth can be seen to "reclaim nature as a resistance strategy and as a means of negotiating" their maternal identities, so that "resistance to structural disadvantages is recoded as resistance to medical control" over their bodies. Thus, "the feminist project is redirected against medicalization instead of, for instance, against patriarchy or capitalism" (LaChance Adams and Lundquist 2013, 209–210).

It is beyond the scope of this chapter to engage the complexities of Johnson's analysis. For the purposes of my Heideggerian interpretation of nontechnological birth, such as Lamaze, Johnson's analysis helps us to see that Lamaze births and medicalized births in a hospital are enframed. In fact, when Johnson reports that privileged Canadian-born women tend to embrace nontechnological childbirth because they see it as a "meaning making" cultural practice, while "immigrant women from developing countries" tend to flock to medicalization because they associate it with safety and good "risk management" (213), regulating and securing birth, our Heideggerian framework allows us to see this as a false dilemma. The two criteria are intimately connected insofar as risk management is an enframing practice that organizes multiple social institutions. Thus, even though "being able to achieve the ideal birth experience is often tied to class or to the ability to have continuous health care from a provider" (LaChance Adams and Lundquist 2013, 234), this privilege is far from a way out of Heideggerian enframing. From this we can infer that women who might eschew reproductive enframing are women who fall wildly outside of mainstream childbirthing norms due to severe social injustices perpetrated against them and/or through their own choices. This is not likely to apply to the Lamaze woman.

Lamaze promises the ideal birth experience by teaching women breathing techniques designed to achieve a painless childbirth. With the Lamaze method women learn(ed) psychological and physical skills for birthing without drugs and medical technologies. This restored the dignity and power they had lost during previous decades and wrested some control of birth from the obstetricians, bringing it back into their own hands, albeit for a short time. Although most women used the Lamaze method in a hospital, a birth center housed within a hospital, or in a freestanding birth center (Michaels 2014, 132)[26] in which medical technology was available if needed, few women actually used the technology. Instead they relied on their acquired skill, trusted their bodies, and worked with their partners to see the birth through.

Fernand Lamaze became acquainted with the psychoprophylactic method, influenced by Pavlovian conditioning, on his trip to Russia in 1951. There he learned that most women labored naturally in hospitals and without the use of drugs that were so prevalent and, in his opinion, harmful in the West, especially in the U.S. That is, he learned that labor was work (Lamaze 1970, 25, 32), something a woman actively and consciously does herself—learning to breathe with a great deal of control and precision, avoiding passivity and relaxation as they increase sensitivity to pain. The goal is to recondition the impulses of the nervous system by changing deeply entrenched attitudes, language, and behavior.[27] A medical *techné* is certainly used, but almost always without technologies. This *techné* gives the pregnant woman a heightened sense of control and mastery over her body. According to Lamaze,

> The Read method is based on the theory of eliminating tensions caused by fear, and thereby letting nature take its course unhampered by harmful emotions. The Pavlov method, while it agrees with the principle of *conquering fear by knowledge*, also makes use of conscious mental and physical control of the birth process. This control is attained through exercises and education designed to build conditioned reflexes which will stand up during the stress of labor and enable the woman to direct her own delivery... That is why we do not call our system natural childbirth. *The final result should be better than nature.* (Lamaze 1970, 27; my emphasis)

According to this Cartesian subject-object model, an anonymous and elusive "nature" is controlled and optimized through a personalized integration of technique rather than being externally regulated by technology.[28] In a Soviet-style masculinist framework, the woman shows up as a laborer, a self-possessed subject. Her mind and body are the instruments of her technique. She has become a subject who is confident in her ability to control her body, to stay on task, and to "regulate and secure" her birthing progress from within. Lamaze describes this *techné* as one that is predicated on the development of "rational knowledge, analysis and control" (Lamaze 1970, 110). As one doctor put it, "Lamaze's emphasis on discipline and mastery are well suited for someone who likes to be in control. The ideal Lamaze woman is, in fact, like a superbly trained athlete who has disciplined herself to perform under intense pressure" (Verny 1981, 110).

The Lamaze woman is caught up in a paradigm of performance and expertise. She plans and trains in order to conquer her fear of pain and ultimately to conquer pain. Instead of simply accepting the suffering that may come with childbirth or seeking refuge in drugs, she now triumphs over the pain through discipline and self-control. The practice of Lamaze illustrates an early stage of reproductive enframing. The woman's work is organized by the desire to order and control the birth process as closely as possible and to avoid the hospital's technocracy. Instead of these norms being imposed on her, she shows up as the subject who *actively imposes them on herself, her body.* Unlike the self-imposition of the woman who opts for the epidural, however, this self-imposition is engaged, skilled, and empowering. The presence of her Lamaze coach and partner underscores her new position as subject in a hospital setting where she used to be a mere object pushed around by doctors.[29]

Yet a closer phenomenological scrutiny reveals this subject to be, in fact, lacking control in important ways. The Lamaze woman has limited control over "nature" going wrong. If things go "wrong" her *techné* is not able to handle the medical complications and emergencies that may arise. If the technique does not work for her, she may have to endure the pain and, on the traditional Lamaze model, be dissuaded by her Lamaze coach from asking for pain relief drugs (Michaels 2014, 125). The Lamaze method often makes for a slow birth process. It is not efficient. Technological devices compromise a woman's *techné* but quicken the process with plenty of precision. Many women today seem to just want birth to be "over with" so they can see the baby. It seems that compromised efficiency and control are chief contributing factors why low-tech and drug-free childbirth continues to wane (Michaels 2014, 143) while the number of technocentric births skyrockets. In fact, in order to survive in a technocratic market in the last couple of decades Lamaze has started to accommodate some technical intervention and has softened its standard critique of natural homebirth as lacking in standards, skill, and commitment (Michaels 2014, 143–154).

Lamaze has thus become more inclusive while also holding on to the precept that achieving a conscious and painless childbirth requires work and engaged agency. This basic tenet serves the following feminist purposes: While the Lamaze approach, as noted above, retains a Cartesian, essentialist conception of the self that has traditionally been harmful to women, the active Lamaze woman can be seen to contribute to the displacement of some longstanding essentialist (mis)conceptions about women in general, and about specific groups of women in particular. The second section above discussed a dominant nineteenth-century patriarchal view that framed bourgeois women in Europe and North America as unable to withstand the pain of childbirth; the women's putative frailty was said to require medical intervention, including sedation. Candace Johnson underscores this view in her essay, "The Political 'Nature' of Pregnancy and Childbirth," in the context of "the myth of painless childbirth." While Lamaze shows that painless childbirth is a matter of individual effort and mastering breathing techniques, previously held dominant views attributed painless childbirth or the ability to bear children "with great ease" and "without any assistance" to "the women of Africa...and indigenous women elsewhere" (LaChance Adams and Lundquist 2013, 206). Painless childbirth did not apply to European women—that is, white women—who were thought to require great assistance and to experience great pain in childbirth. Since this pain was associated with "a higher level of human development" and with a "delicate constitution" that was deemed to be simply "natural" (LaChance Adams and Lundquist 2013, 207), it is easy to see that the script of painless childbirth was used to perpetuate a racist dualism according to which white women were considered superior to nonwhite women, who were deemed naturally inferior. The supposed proximity of the latter to nature translated into their automatic affinity for all things physical and biological and an automatic distance from intellectual work, the rational thought that was the marker of "the higher level of human development."

This patriarchal pitting of white women against nonwhite women is abstract and benefits white women as little as it benefits African women and indigenous women, since none of the women are recognized as doing meaningful work or wielding any power in childbirth. While white women are trapped in their "delicate constitutions" and are pushed around as objects by male doctors, nonwhite women's "ease" in childbirth is not respected or associated with indigenous knowledge and/or skill. The practice of Lamaze, by contrast, presents a feminist alternative. It holds on to the notion that childbirth can be painless but rejects any mystification or orientalization of this event. In this sense, it can be seen as a pluralistic arena in which individual women can reject oppressive norms and affirm the value of childbirth. Despite Lamaze's promise of an empowered and painless birth, however, it continues to be marginalized by its technocratic counterpart and external forms of medical control. I have suggested that the main reasons for this displacement have been inefficiency and a restriction of control. Because traditional Lamaze strives for maximum control and optimization—without advocating slowness but merely accepting it as a possible side effect—it helps to disclose the enframing that constitutes our technological world.

Conclusion

In recent decades, feminists have criticized growing technological intervention in childbirth. Many allege that it is another form of patriarchal oppression of women's bodies and agency for which the appropriate antidote is a wholesale shunning of reproductive technology and a return to natural, drug-free childbirth.[30] While I recognize the legitimacy of this concern, I suggest that both technological and low-tech or no-tech options, like Lamaze, presuppose and promote order(ability) and control. With its emphasis on restoring women's dignity and acquiring skills that will help the mind to control the pains of the body, traditional Lamaze reflects unquestioned Cartesian subject-object assumptions. The Lamaze woman's self-imposition of a training regimen that will order and optimize her body, however, brings in view the collapse of the subject-object distinction that is characteristic of Heideggerian enframing. This collapse is more visible in the typical (technophilic) hospital birth, in which women's reproductive bodies are harnessed by a variety of technologies intended to produce an ordered and orderly birth. The low-tech/no-tech instrumental attitude of the Lamaze woman, directed at her body, already helps paradoxically to provide a glimpse into this collapse. By combining Heidegger's theory of technology with feminist phenomenology, I have shown that when choosing between technological and nontechnological (low-tech) birth pregnant women may be choosing the same fundamental approach: order(ability), control, and optimization. Thus, what is essentially one option from the point of view of enframing masquerades deceptively as widely divergent, even diametric, choices. Once we begin to understand how reproductive enframing influences women's birthing choices, however, we may come one step closer to developing a free relationship to birthing technologies.

NOTES

1. This sense of history is ontological and it refers to what Heidegger calls, the history of being, rather than to a sociological and empirical account of events produced by human beings. Heidegger calls the latter historiography. In the ontological sense, history refers to discrete horizons of meaning. The first one is articulated in Plato's metaphysics. We inherit these horizons of meaning and they serve as the "measure" or the conditions of the possibility of our self-understanding and relationship to the world. See Heidegger, "Science and Reflection," in *The Question Concerning Technology and Other Essays*, 175. Heidegger writes, "The word *Historie* [historiography] means to explore and make visible, and therefore names a kind of representing. In contrast, the word *Geschichte* [history] means that which takes its course inasmuch as it is prepared and disposed in such and such a way, i.e., set in order and sent forth, destined."

2. Andrew Mitchell translates *Ge-stell* as positionality in his essay "Positionality," in *Bremen and Freiburg Lectures: Insight into That Which Is and Basic Principles of Thinking* (Bloomington: Indiana University Press, 2012).
3. A few disparate examples include parental intrusion into a child's learning curve for the sake of accelerating the child's cognitive skill set at an early age, drawing up a birth plan to optimize the woman's birthing experience, and binge exercising.
4. According to Carolyn Merchant, "beginning with the seventeenth century, mechanics or 'the science of matter in motion' could be used to describe the entire universe—the human body, the physical surroundings, and the larger cosmos" See *The Death of Nature*, 1989, ch. 7). It may seem strange to challenge the assumption that matter cannot think. Yet there are historical and metaphysical precedents, according to which, matter or nature is viewed as *ensouled* and seen to possess a kind of intelligence.
5. Merchant, *The Death of Nature*, 204.
6. These other values, i.e., tranquility, piety, cooperation, etc. can be seen as forming a contrast class that enables a glimpse into the historical contingency of the dominance of current technological values. On a Heideggerian reading the recentering of the contrast class "values" could signal the dawn of a post-technological epoch.
7. For a detailed account of Heidegger's problematic foreclosure of ontological transformation through human means, see Dana S. Belu and Andrew Feenberg, "Heidegger's Aporetic Ontology of Technology," *Inquiry* 53, no. 1 (2010): 10–19.
8. See Lorraine Markotic, "Patriarchy, Enframing and the New Revealing: O'Brien's Philosophy of Reproduction and Heidegger's Critique of Technology," *Hypatia* 31, no. 1 (2016): 130.
9. Foremost among feminist work inspired by Heidegger's interpretation of *physis* is Luce Irigaray's *The Forgetting of Air in Martin Heidegger* (1999). It does not, however, focus on reproductive technology or childbirth. An abstract account of birth as natality, that is not necessarily feminist, is available in Anne O'Byrne's *Natality and Finitude* (2010). The most sustained Heideggerian and feminist account to date of the use of reproductive technologies is Maren Klawitter's article "Using Arendt and Heidegger to Consider Feminist Thinking on Women and Reproductive/Infertility Technologies" (1990). The article focuses mostly on reproductive technologies used during conception and surrogacy.
10. The main focus of this chapter is on the phenomenology of childbirth. While it touches on some ethical and political issues regarding the medicalization of women's bodies in childbirth, a more in-depth engagement with issues related to reproductive justice fall outside the scope of this chapter. For an

excellent account of reproductive justice see Dorothy Roberts, *Killing the Black Body: Race, Reproduction and the Meaning of Liberty* (1999).
11. Elective cesareans can be seen to bring the reproductive enframing into focus. They illustrate a heightened desire for order, control, and efficiency. The pregnant woman can satisfy this desire with the aid of medical technology. As is well known, elective cesareans are scheduled far in advance so the woman knows exactly when the baby will be delivered and thus she can avoid the inconvenience of waiting for labor to start, painful contractions and tearing during the actual birth. Elective cesarean are presented as a rational option, a sign of medical progress.
12. See Louden, *Death in Childbirth* (2002), 297. Anecdotal evidence suggests that the American and European technophilia seems to have included remote areas of Eastern Europe. According to my beloved grandmother, Viorica, giving birth in hospitals during the 1940s was seen as modern, civilized and progressive. She birthed my mother at home, in 1947, in the small village of Tiream in Transylvania, Romania, because no hospital was located nearby. Once she went into labor, she simply called out to the village midwife across the fields. The midwife helped her give birth "the old fashioned way" and then fed her spicy, homemade sausage and a shot of tsuica (Romanian eau de vie) to help with the recovery from the long labor and birth.
13. Stephen Wilkinson, in *Choosing Tomorrow's Children: The Ethics of Selective Reproduction* (2010), presents one of the standard bioethical definitions of enhancement, that is, "The Non-Disease-Avoidance-Account" as referring to "any improvement through modification of selection that goes beyond, or is something other than the avoidance of disease" (187–188).
14. One might reply that this is how "good" technologies develop, through trial and error. However, a feminist retort would point out that, in this case, the cost is too high.
15. See J. Drife, "The Start of Life: A History of Obstetrics," in *Postgraduate Medical Journal* 78 (2002): 311–315; doi:10.1136/pmj.78.919.311.
16. In 1882 the German gynecologist Max Sänger invented the modern cesarean technique. His breakthrough was to suture the uterine cut after the extraction of the fetus. Before Sänger's technique, the cut was not closed, leaving many women to die from severe hemorrhage and sepsis. Sänger's procedure reduced maternal mortality rates. Coupled with the use of chloroform the suturing technique increased the number of subsequent cesareans and augmented the social perception, soon to become dominant in the US, that birthing in hospitals is safer and more progressive than home birthing. It consolidated physician-controlled birth and constituted a new frontier in obstetric intervention.
17. W. Gogarten and H. Van Aken, "A Century of Birth," *The International Anesthesia Society* 91 (2000): 772. An earlier version of the spinal-epidural anesthetic was available in the early 1900s but the lack of monitoring and

skill in administration resulted in a high mortality rate and this analgesic method was abandoned for a while, landing obstetrics in what was considered to be the medical dark ages.
18. This increase could be related to physicians practicing "defensive medicine." See S. Burrow, "On the Cutting Edge..." (2012).
19. Reproductive enframing applies Heidegger's concept of the enframing to the domain of reproduction. It refers to interconnected and mechanized (medical) processes during conception, pregnancy, and birth, which reveal women and doctors as resources for a medical network that privileges its own optimization. For more, see Chapter 3 above.
20. Many feminists are critical of the patriarchal foisting of reproductive technologies on women. Radical feminists, for example, have voiced sharp criticism of the harms associated with the use of ARTs. They see these as (re)new(ed) forms of a patriarchal oppression of women.
21. Adrienne Rich warns in *Of Woman Born: Motherhood as Experience and Institution* against the patriarchal power inherent in the use of technologies that appear to help women. She cautions that "freedom from pain," like "sexual liberation" places a woman physically at men's disposal, though still estranged from the potentialities of her own body. While in no way altering her subjection, it can be advertised as a progressive development" (171).
22. Compare this type of objectification with Sandra Lee Bartky's account of sexual objectification in her essay "On Psychological Oppression" in *Femininity and Domination: Studies in The Phenomenology of Oppression* (New York: Routledge Press, 1990), 27.
23. According to recent statistics compiled by the CDC, 99 percent of women gave birth in hospitals in the U.S. in 2006. Out of these only 7.5 percent used midwives; the other 91.5 percent were delivered by physicians with the aid of technology. Of course it is possible that technology was also used in midwife-assisted births but this is usually uncommon. See www.cdc.gov 57, no. 7, p. 16. I have not been able to find statistics regarding out of hospital, nontechnologically mediated births. In light of the reproductive enframing, this is a striking absence as it highlights the types of birth that matter—i.e., efficient, medicalized, and monitored births—and the others that do not matter, nontechnological births.
24. While the rise of the women's movement and its sociopolitical criticism of patriarchy opened the doors to natural birth it is important to note that Lamaze in the U.S. (ASPO), especially its founder Elizabeth Bing, was extremely critical of natural birth because it was perceived to lack skill and commitment. This position did not soften until the 1980s. Like natural birth, Lamaze was practiced in birth clinics. Birth clinics opened throughout the country in the early 1970s. In these lightly medicalized spaces women were attended by experienced midwives and were allowed alternatives to hospital births.

25. For a distinction between medical assistance and medicalization see Ann Garry, "Medicine and Medicalization: A Response to Purdy," in *Bioethics* 15, no. 3 (2001): 262–269.
26. "Birth centers offer a low-tech, comfortable place for childbirth that's safer than having your baby at home if problems arise. At an accredited birth center you will be cared for by licensed professionals, usually a midwife and a nurse, with a backup hospital nearby and a doctor on call in case of an emergency. Typically, a birth center is an independent facility, though a growing number are affiliated with and often housed inside hospitals."
27. Ibid., 104–107. It is beyond the scope of this chapter to go into further physiological detail but suffice it to say that following Pavlov, Lamaze reinforced the connection between the cerebro-spinal system "responsible for voluntary movement and sensation" and the visceral nervous system that "governs our internal organs." Since the brain controls both systems, by controlling the information processed by the brain we can modulate the signals that the brain sends out to the rest of the body and its organs.
28. Nonetheless, since Lamaze was mostly practiced in hospital settings technological support was available if needed.
29. The role of the Lamaze coach is controversial. According to some accounts, traditional coaches exerted tight authority over the birthing process, demanding obedience and even shaming women if they screamed in pain or asked for drugs to ease their pain. Today, under Lamaze International, the guidelines are much softer. See Michaels (2014), 125, 114–141.
30. After the birth of the first IVF baby (a girl) in England in 1978, a large body of feminist literature emerged that criticized technological intervention in the womb as patriarchal and technology itself as patriarchal. It exposed the abuses of women by the medical industry. See Gena Corea, *The Mother Machine: Reproductive Technologies from Artificial Insemination to the Artificial Womb* (New York: Harper & Row, 1985); the radical feminist perspectives expressed in Patricia Spallone and Deborah Lynn Steinberg, eds., *Made to Order: The Myth of Reproductive and Genetic Progress* (Oxford: Pergamon, 1989); Rita Arditti, Renate Duelin Klein, and Shelley Minden, eds., *Test Tube Women* (London: Pandora, 1984); and Emily Martin, *The Woman in the Body: A Cultural Analysis of Reproduction* (Boston: Beacon, 2001). Gena Corea recounts that Lesley Brown, the mother of Louise Brown, the first test tube baby, assumed that she was one in a series of women to undergo IVF. The doctors did not inform her that she was the first successful experiment. She was not informed about the risks of the procedure and was terribly scared when she experienced sudden and profuse bleeding as a result of the IVF procedure. Lesley Brown experienced this bleeding while riding the subway.

CHAPTER 6

The *Poiésis* of Birth

Abstract This chapter explores an alternative to enframed birth. This alternative of birth as a bringing-forth connects the later Heidegger's interpretation of *poiésis* with Sara Ruddick's feminist account of a "work of conscience" (a concept that seems to owe an unacknowledged debt to Heidegger's *Being and Time*). I argue that childbirth is exemplary for bringing-forth and that water birth is exemplary for childbirth. My account of birth as a bringing-forth transposes aspects of Ruddick's account of a mother's work of conscience, directed at her children, onto a pregnant woman's relationship to herself and to her fetus. Laboring and birthing as a bringing-forth require that a woman engage in the reflective practice and struggle that are necessary to resist dominant forms of birthing that treat mother and child as mere resources. I argue that empathy is necessary for birth.

Keywords Bringing-forth · Work of conscience · Authenticity · Inauthenticity · Empathy

I begin this concluding chapter by underscoring the tension between the two phenomenological interpretations of Heidegger's concept of enframing—total enframing and partial enframing—the paradox discussed in Chapter 2. According to total enframing, any awareness of the enframing is blocked, while according to partial enframing, an awareness is introduced that compromises the totality of the enframing. This awareness contradicts the totality that is central to Heidegger's view of *Ge-stell* as a forgetting of the clearing of being. In Chapter 5 I presented a feminist phenomenology of childbirth according to which forms of technophilic and technophobic birth are unwittingly stuck in the reproductive enframing. This underscores the totalizing account of enframing, yet also paradoxically invokes partial enframing, since it presents an awareness of enframed birth *as* enframed.

While in the previous chapters I focused on the ways that women's reproductive bodies are challenged-forth, in this chapter I explore a possible alternative to enframed birth. Childbirth as a bringing-forth can be seen as exemplary, I argue, for Heidegger's interpretation of *poiēsis*. Given the extent of Heidegger's attention to this ancient Greek concept, it is surprising that he never connects it to childbirth. My vision of nonenframed childbirth as a bringing-forth connects Heidegger's interpretation of *poiēsis* with Sara Ruddick's feminist account of a "work of conscience." Birth as a bringing-forth transposes aspects of Ruddick's account of a mother's work of conscience, directed toward her children, onto a pregnant woman's relationship to herself and her fetus. I argue that today labor and birth as a bringing-forth require that a woman engage in the reflective practice and struggle that are necessary to resist dominant forms of birthing that treat mother and child as mere resources. Attentive love, empathy, and proper self-trust aid this struggle. I argue that empathy is necessary for the experience of birth as a bringing-forth and show that water birth renders empathy in childbirth highly visible.

MATERNAL AGENCY AS A WORK OF CONSCIENCE

In *Maternal Thinking: Toward a Politics of Peace* (1990), Sara Ruddick claims that maternal thinking requires preservative love, fostering the growth and training of young children. Ruddick asks whether it is possible for mothers to approach this training authentically—educating children as a "work of conscience"—rather than inauthentically by dominating them. Ruddick's dualistic account of inauthentic/authentic training appears to tacitly appropriate into a normative feminist framework aspects of Martin Heidegger's phenomenological ontology. In *Being and Time* (1927), Heidegger draws an ontological distinction between two ways of caring that he famously describes as "leaping-in" and "leaping-ahead." Section 26 of *Being and Time* distinguishes the two ontological ways or "two extreme possibilities" of caring for another. The first, "leaping-in" (*einspringen*), is oppressive and controlling, though it may appear in the guise of helpfulness.[1] Heidegger writes:

> Concern takes over what is to be taken care of for the other. The other is thus displaced, he steps back so that afterwards, when the matter has been attended to, he can take it over as something finished and available or disburden himself of it completely. In this concern, *the other can become one who is dependent and dominated, even if this domination is a tacit one and remains hidden from him.* (Heidegger 1996, 114; my emphasis)

The second kind of caring, "leaping-ahead" (*vorspringen*) is distinguished from the first as

> the possibility of a concern which does not so much leap in for the other as *leap ahead of him* not in order to take "care" away from him, but to first give it back to him as such. This concern which essentially pertains to *authentic care*; that is, to the existence of the other, and not to a what which it takes care of, helps the other to become transparent to himself in his care and free for it. (Heidegger 1996, 115; my emphasis)

Heidegger adds that between these two possibilities there are many other "mixed forms," and Ruddick, in her analysis of maternal inauthenticity, acquiesces that it "is an attitude that admits of degrees" (Ruddick 1990, 113).

When reproductive technologies "leap-in" for the birthing woman they often replace birth with surgical extraction. This realization can open up the possibility of preserving birth. The task is to recognize the great hold that leaping-in, as the effort to "still the flux"[2] of birth, exerts on women and on medical authorities. With its tendency to impose, order, and dominate under the guise of being helpful and efficient, leaping-in points to central features of reproductive enframing. Heidegger characterizes leaping-in as an inauthentic mode of being-with (*Mitda-sein*). Transposing his analysis to the context of technologized birth, we notice the different forms of medicalized leaping-in. Doctors can be seen to leap-in when they impose fetal monitoring technology, seduce with epidurals, subject the birthing woman to a stringent timeline, and/or enforce cesarian sections.[3] As I argue in Chapter 5, the overuse of these medical interventions tends to transform birth into a mechanized delivery system run by a medical team that lacks empathy for the woman and marginalizes her agency. In sum, to borrow

Lauren Freeman's words, "*leaping-in* for another blinds us to the existence of the other as *Dasein*; it pushes her out from her place, job, or role, and in so doing, takes the matter at hand over from her... This is another way of saying that the other becomes unfree."[4]

Overwhelmed by the anticipation of pain or by the pain of real contractions, the birthing woman may "disburden" herself of her agency as she hands it over to the doctor. She becomes "dependent and dominated," yet "this domination is a tacit one and remains hidden" (Heidegger 1996, 114). Where does it hide? In the belief that the interests and well-being of those it dominates—that is, above all the woman herself—are actually being served. Doctors and nurses, for instance, are on call to expedite labor and to make birth efficient and painless. As Ruddick points out, however, birth "cannot be rescued... by denying its pain. Rather, birth's distinctive conjunction of erotic excitement, physical pain, and social promise can provoke reflection on the place of pain in life" (Ruddick 1990, 214). Pain in childbirth ought to be distinguished from injury or harm. While pain is usually temporary and a part of life, injury tends to induce a relatively more persistent form of pain. Excessive avoidance of pain can produce injury. The pregnant woman can choose to forego all pain in birth by scheduling an elective caesarean. While this procedure manages the pain of giving birth, it also deletes the experience of birth, reducing the pregnant woman to a marginal character in her own story. The medical machinery "leaps-in" and discloses her "as a thing, or an implement that can be controlled, mastered, dominated, ignored, or humiliated," again to borrow Freeman's words.[5]

Let me underscore that this "leaping-in" requires the willing participation of the woman. Birthing women often solicit or easily agree to a host of *unnecessary* medical interventions. Ruddick uses the term *inauthenticity* to identify the willing abdication of maternal thinking in favor of external (parental) expertise (Ruddick 1990, 111–113). Maternal inauthenticity describes mothers inviting others to "leap-in" as experts and authority figures in the training of their young children. In childbirth it shows up when women too easily abdicate their know-how, power, and responsibility to medical experts. Ruddick describes maternal inauthenticity as follows:

> Abdication can take many forms... The abdicating mother talks as if the authorities are legitimate, as if their will should be obeyed... Fear of the gaze of others can be expressed intellectually as "inauthenticity," a repudiation of one's own perceptions and values... *It is when the struggle is denied*

and rendered invisible that [maternal] thinking becomes inauthentic. (Ruddick 1990, 112–114; my emphasis)

In choosing a medicalized birth, especially one afforded by elective cesareans, women can be seen to abdicate the struggle and commitment to birthing. When experts take over the birth they conceal the individual woman's own possibilities, even as this concealment is dismissed or covered over. This feminist appropriation of Heidegger reflects the slip from what he calls the "average, everyday" and public understanding, which "determines what and how one 'sees'" (Heidegger 1996, 159), into inauthenticity, a way of conferring absolute authority upon this commonplace "seeing."

Ruddick claims that inauthenticity is ultimately about "form, not content" (Ruddick 1990, 113). This suggests, along unacknowledged Heideggerian lines, that there is no particular maternal choice or action that is *a priori* inauthentic. Performing a quasi-ontological move, Ruddick's "form" refers to any preemptive abdication and disburdenment of commitment or responsibility coupled with a ready-made glorification of medical authority. She continues:

> It is not when they submit or are prudent or timid that mothers are inauthentic. It is when they lose sight of the cost of prudence, deny their timidity, and tell their children that unquestioning obedience is actually right... When she thinks inauthentically a mother valorizes the judgment of dominant authorities, letting them identify virtues and appropriate her children for tasks of their devising. (Ruddick 1990, 113)

Likewise, when pregnant women automatically valorize the judgment of dominant medical authorities without conscientiously questioning their prescriptions or asking what kind of birth may be right for them and their unborn child, they are abdicating the struggle, allowing a most intimate experience to be appropriated by experts. The aim of this struggle is to have an individuated birth and to promote the actualization of maternal values in social and medical institutions.

According to Ruddick, the "work of conscience" ought to be a corrective to maternal inauthenticity. The central feature is a vigilant moral *struggle* to build patience, reflectiveness, trust, love, and "a capacity for guilt and shame" (Ruddick 1990, 109). Ruddick does not adequately explain this capacity. She does insist, however, that a mother must stay attuned to her own value system and practice "conscientiousness—the ability to identify, reflect on, and respect the demands of conscience" (Ruddick 1990, 109). While Ruddick claims that these cognitive capacities and virtues are motivated and directed toward improving the education of young children, they can be fruitfully extended to the education that a pregnant woman undertakes on her own behalf. Shifting the focus from the welfare of the child to the pregnant woman removes some of the patriarchal baggage from Ruddick's account[6]—for according to Western patriarchal traditions, the principal goal of a woman's life is to become a mother, and the principal goal of being a mother is to take care of her children and not of herself.

Ruddick's account of the work of conscience also bears a traditionally virile and masculinist imprint that is not questioned. For instance, "to identify proper trust as a virtue is not to identify an achievement but an ongoing and difficult struggle" (Ruddick 1990, 119), and when mothers "identify proper trust as a virtue and attempt the discipline of attentive love, all that they can assure is that the work of training will not become a battlefield but a hard, uncertain, exhausting, and also often exhilarating work of conscience" (Ruddick 1990, 123). The work is hard because mothers, and pregnant women even more so, are "prey to self-loss . . . indifference, passivity, inquisitorial scrutiny, domination, intrusiveness, caprice, self-denial and self-protective cheeriness" (Ruddick 1990, 120). Although Ruddick insists that maternal thinking is an ambivalent struggle strained by various emotions, impulses, virtues and vices, rather than a triumph on the battlefield, this seems to be a distinction without a difference. The price of failing in the struggle for authenticity is high. Children are injured and lose trust in their mothers. As on a battlefield, this trust must then be fought for and won back.

Ruddick's account of maternal conscience seeks to empower the woman who has become a mother and to underscore her confidence in herself, but it can also seem critical and burdensome to the mother. Lest she become inauthentic, the mother must remain constantly alert, vigilant, and morally supple in the education of her children. She supports them but no one supports her. In fact, during this unceasing work of conscience she must support herself. Ruddick's account seems to repeat the patriarchal gesture of overloading the mother with responsibility for others, however it also succeeds in shifting the focus from a patriarchal objectification of mothering to an empowering maternal subjectification that can be helpful for a nonenframed childbirth.

Ruddick's maternal value system can be seen to hold up the interconnection between the virtues of reflective judgment, attentive love, empathy, and self-trust for women who desire to birth their children. Together these virtues constitute a sense of maternal agency that extends back before the inception of labor. As stated earlier, the questioning of medical authority, especially when that authority drives to impose a medicalized birth, practices what Ruddick calls conscientiousness and reflective judgment.[7] When a woman focuses reflective judgment and attentive love on her own labor and childbirth, she is acting empathetically toward herself and toward her fetus. Drawing on the ethical work of Simone Weil and Iris Murdoch, Ruddick defines attentive love as "a kind of knowing that takes truthfulness as its aim but makes truth serve lovingly the person known" (Ruddick 1990, 120). Maternal truths that would not reveal themselves otherwise show themselves to the "patient eye of love." Yet attentive love is threatened on all sides by the lure of fantasy, a cognitive escape that pregnant women often fall prey to. Fantasy disguises itself as a "reverie designed to protect the psyche from pain, self-induced blindness designed to protect it from insight" (Ruddick 1990, 120).

In addition to reflective judgment and attentive love, proper trust is an essential part of maternal authenticity. It "is a virtue of which unquestioning obedience or blind trust are degenerative forms" and it requires "clear judgment that does not give way to obedience or denial" (Ruddick 1990, 118–119). In the previous chapter we saw an example of self-trust in the analysis of the Lamaze laborer who relies on her acquired skills to give birth, yet does not deny medical intervention if it is needed. The significant corrective that Ruddick's work of conscience can be seen to bring to Lamaze birthing is a critical awareness about domination. A pregnant woman's tendency to self-dominate is checked by reflective judgment and empathy.

While Ruddick cautions against external forms of domination, these forms of domination are internalized by the Lamaze laborer. As I explain in the previous chapter, the Lamaze laborer learns and trains to dominate herself, thereby transforming birth into a performance. This aggressive and masculinist approach in which the woman exercises control from within is a liberating step over and above the control exerted from without by medical authorities, however it also transfers the reproductive enframing to the agent. In her quest to dominate nature, to erase pain,[8] the Lamaze laborer can be seen as "leaping-in" for herself.

Ruddick's account is cautious about domination and can be useful for envisioning childbirth as a woman-centered alternative to Lamaze. In this alternative scenario a woman is invited to practice the "clear judgment" that some pain may be an acceptable part of the birthing process and that it need not translate into harm or permanent injury. This eschews the escape into fantasy. The notion that all pain is "bad" and should be avoided reproduces paternalistic interpretations of women's "gentle nature." This paternalism tends to play into the maternal feelings of fear and anxiety that are usually associated with childbirth. A work of conscience includes learning to accept these feelings without pushing them away by means of training or technology. It helps to prepare for the flux of birth, while recognizing that no amount of preparation can organize the flux. It sees birth as a unique and intense encounter with life without attempting to stave off this life-flux. Finally, a work of conscience helps to reveal the foundational role of empathy in birth as a bringing-forth.

Empathy in Childbirth as a Bringing-Forth

In an apparent criticism of Heidegger's well-known account of throwness (*Geworfenheit*) in *Being and Time* (Heidegger 1996, 192), Ruddick comments that too much "groundbreaking" philosophical discussion has been focused on being "thrown" into the world and not nearly enough philosophical attention has been devoted to birthing (Ruddick 1990, 192). This oversight is due to a patriarchal privileging of reason over the body, and especially the birthing body. While Ruddick is correct that throwness does not focus on the birthing body, Heidegger's phenomenology of throwness, like Ruddick's feminist analysis, undermines the primacy of reason for making sense of our world. Because throwness is an ontological structure that shows *Dasein* as always already absorbed in a world of cares, concerns, and projects that it must take over or ignore, it is actually compatible with Ruddick's critique of reason. Neither critique is anti-rational, but critical of totalizing, representational, and calculative forms of rationality. Ruddick writes:

> The compulsion to control and minimize birth was expressed in the ideals of reason that dominate Western philosophy...A rational person is one for whom the capacities and values associated with reason, control and order properly subordinate capacities represented by the body...In opposing reason and masculinity against the body and femininity, it is *birthing labor* and what it represents [bodily mortality] against which reason sets itself.[9]

Rational thinking sets itself up as other than the birthing body with its "irregularities, pain, vulnerabilities and decay" (Ruddick 1990, 196). The truth of reason is built on their exclusion. Ruddick claims that "since the relationship of birth is taken to exemplify merging or the failure of individuation, reason would require and promise autonomy, which in turn would be defined in terms of separation and detachment... Reasonable people would be expected to provide and count on stability and regularity."[10] This is the promise of a medicalized birth.

As Ruddick incisively remarks, however, "birth is a beginning whose end and shape can be neither predicted nor controlled"[11] lest it cease to be birth and become delivery. The embodiment of reason in the use of advanced reproductive technology is geared toward occluding birth as the uncontrollable. Since it is women who birth, the history of "reason" can be seen to inflict a sexist synecdoche upon women, reducing them to their womb and silencing their speech (*logos*). This problem is compounded when women act inauthentically and silence themselves. If "Western conceptions of what it is to be reasonable are intertwined with a fear and resentment of birthing female bodies" (Ruddick 1990, 195), then it follows that concealing birthing bodies makes good sense. Hiding these swollen, incontinent, and irregular bodies in hospital gowns, behind partitions and technological machinery, further strengthens the power of a medical reason that is already "idealized as active, autonomous, controlling, progressive and socially powerful"—and already undervalues the birthing bodies.

No one can accurately anticipate the start and duration of labor. It is still largely unknown how labor starts, and some scientists suspect that the fetus triggers the process. Approaching childbirth as a bringing-forth rather than a challenging-forth requires some surrender to contingency and openness to the unknown. It may also require that we see birth, at least in part, as "a beginning whose origins are mysteriously concealed." (Bigwood 1995, 38) While labor demands relaxation and openness it also demands attention and active maternal participation. The mere prospect of these seemingly incompatible demands—to stay relaxed *and* focused—provokes anxiety, and this often invites maternal flight into medical intervention. The anxiety is not necessarily about any one particular obstacle, rather it appears pervasive and diffuse: it hovers over new birth like a specter. Often the unspoken fear of fetal death provokes this anxiety.[12] The anxiety can also gather around a woman's concern for her own physical survival and/or the survival of her prematernal self, the loss of her freedoms. This idea is underscored by

Simone de Beauvoir in her phenomenology of pregnancy and birth in *The Second Sex:* she claims that for a woman who approaches childbirth, "every transition is fraught with anxiety; childbirth appears especially terrifying. When the woman approaches her term, all her childish terrors come to life again."[13] Avoiding pain, fear, and anxiety about one's own death, biological or symbolic, along with fear of injury to the fetus, are central reasons for relying on reproductive technologies that recast an unknown process into a secure medical intervention. For the most part, however, they also preclude the possibility of bringing-forth one's child.

Childbirth can be seen as the paradigmatic event of "bringing-forth." It may be an acute case of patriarchal blindness on Heidegger's part that despite his numerous and far-reaching commentaries on bringing-forth he never connects it to childbirth. In "The Question Concerning Technology" he contrasts the *enframed epoch(é)* with that of *poiésis* or "bringing-forth" in the time of the pre-Socratics, writing: "Bringing-forth brings hither out of concealment forth into unconcealment. Bringing-forth comes to pass only insofar as something concealed comes into unconcealment" (Heidegger 1977, 11). This unconcealment puts the "modes of occasioning, the four causes" into play. Some translators warn that the word choice *occasioning* to translate the German word *Veranlassen* "proceeds in the direction of effecting, which is exactly what must be avoided"[14] for a proper understanding of *bringing-forth*. Replacing the translation "occasioning" with "active letting" brings out the nurturing dimension of nature as a bringing-forth. Heidegger cautions us to remember:

> Not only handcraft manufacture, not only artistic and poetical bringing into appearance... is a bringing-forth, *poiésis. Physis*, also the arising of something from out of itself, is a bringing-forth, *poiésis. Physis* is indeed *poiésis* in the highest sense. For what presences by means of *physis* has the bursting open belonging to bringing-forth, e.g. the bursting of a blossom into bloom, in itself (*en heautói*). In contrast, what is brought forth by the artisan or the artist, e.g., the silver chalice, has the bursting open belonging to bringing-forth not in itself, but in another (*en allói*), in the craftsman or artist. (Heidegger 1977, 10–11)

In childbirth, especially in water birth, we see *physis* as *poiésis*, albeit in an ambiguous way. Quite literally the baby emerges into unconcealment. She arises and brings herself forth into the world. At the same time, the mother actively helps to bring the baby forth through attention and empathy, watching over the birth with "the patient eye of love" (Ruddick 1990, 121), nurturing the labor and releasing herself into it. It is unclear if "the bursting forth" belongs to the mother or to the baby or to both of them. The mother must be attuned to labor in an "abetting" way. As Richard Rojcewicz's interpretation of bringing-forth in *The Gods and Technology* points out, abetting is not at all the same as efficiently producing a result, efficiently causing the baby to come out. Rather, it

> is not an efficient cause, where all the agency lies on the one side and all the passivity lies on the other side... By itself, abetting is nothing. That is, it is nothing to one who cannot respond to the abetting... For there to be abetting, there must be activity on the part of both the abettor and the abetted. Likewise, there must be passivity on both sides.[15]

Enduring labor is an ongoing abetting and not a skilled triumph over one's body or a resignation to technological handling. It is a deep attunement that turns on empathy with the baby.[16]

Water birth can be seen to hold up this attunement and empathy as a capacity to feel-with the fetus and with different aspects of oneself. Although empathy is often covered up by an overuse of reproductive technology it is always already inscribed in the process of childbirth. Thus the issue is not whether to choose a birth that is either empathetic or not empathetic, but rather the issue is one of recognizing that empathy is always already a constitutive element in birth that we can choose to either honor or diminish. Heidegger's pithy account of empathy in *Being and Time* helps to bring out this point. He claims that empathy, or what I call a feeling-with, describes a derivative (*existentiell*) way of being with one another predicated on the primordial and neutral existential structure of being-with.[17] He writes, "Empathy does not first constitute being-with, but is first possible on its basis, and is motivated by the prevailing modes of being-with in their inevitability" (Heidegger 1996, 117). Empathy as a feeling-with lies at the heart of birth as a bringing-forth. Feeling-with closely resembles Ruddick's account of empathy as the "ability to suffer or celebrate with another as if in the other's experience you know and find yourself," but without substituting yourself for the other (Ruddick 1990, 121). At first, water birth may seem like a wild idea or a fantasy. But in view of Ruddick's account of fantasy as an escape or "reverie" designed to blunt pain, blind women to insight, and control life "with an abstract plan" (Ruddick 1990, 121), water birth is the sober counterpart to the fantasy of medicalized birth.

Immersed in a big tub of water at home or in the sea, the woman is free to move around. The buoyancy afforded by the water helps to support her weight and reduces the pain of contractions. The woman is not challenged-forth nor does she challenge herself forth through technology or the need to perform. She prepares for the pains of contractions and also welcomes their amelioration by the water. Like the Lamaze woman, she says *no* to technical intervention, but the presence of a midwife signals her willingness to also say *yes* if the need arises.[18] No one "leaps-in" for the woman, nor does she "leap-in" for herself. She is not pushed around by medical staff and authorities. The midwife encourages the woman's labor without intrusion or domination. The woman's "care" is given back to her as she becomes "free for it" (Heidegger 1996, 115), guiding the birth, leading forth a new life. She engages in attentive love and also "understands herself as free...her world is opened up to her and she is not prevented from determining and acting upon her projects."[19] The woman's partner is often in the water as well, participating in the experience rather than merely recording the event from the sidelines, as in hospital births.

The mother and the attending midwife "leap-ahead" of the baby instead of "leaping-in." The fetus is treated empathetically, not in a functional way as a product or a thing. Unlike in many hospital births, where the newborn is yanked out and slapped into breathing, then separated from the mother, wiped, weighed, and rendered immobile by swaddles, in water birth the newborn is recognized as a gentle being. Her coming to life is treated with attentive love. Her first exposure to love and safety happens in the immediate act of being held. Water birth allows for continuous holding of the newborn and this teaches her love and trust. By contrast, some hospitals today charge mothers a fee to have skin to skin contact with their newborns.[20] Thus, we can see how water birth is an example of birth as a bringing-forth that materializes the value of empathy. I am not arguing that water birth is the only way for a woman to bring her baby forth. I am using the example of water birth, however, to show that empathy is an essential component of birth as a bringing-forth, and that this is especially visible in water birth.

When empathy is missing, informed consent fills the vacuum left behind. In light of Heidegger's and Ruddick's accounts of empathy, I see informed consent as a derivative or degenerative form of empathy, a sign that proper trust between a woman and her doctor is missing. The feminist debate about informed consent seems polarizing and further distracts from the issue of empathy. Informed consent requires full disclosure and fair representation of all potential medical, social, and emotional outcomes and risks involved in a medical situation.[21] It includes among its characteristics "competence, voluntariness, the disclosure of information about the diagnosis and therapy, its risks, benefits and alternatives; and comprehension of such information."[22] The definition assumes that this disclosure is possible and that relevant risks are either known or can become known on demand but often this is not the case.[23] It also assumes that the relationship between the medical professional and the patient is based on rational communication.

Liberal feminists view informed consent through a *technophilic* lens, seeing in it a means of respecting women's individual rights.[24] They claim that too much emphasis on the welfare of women as a group sacrifices the freedom of the individual woman to exercise bodily ownership, including the buying and selling of reproductive services.[25] On the other hand, radical feminists like Rowland argue that informed consent promotes patriarchal oppression of women's bodies, and that "full informed consent is not possible considering the [patriarchal] social context within which the medical profession operates."[26] According to some radical feminists, technology is the expression of patriarchal rule in the West and thus it is predicated on a particular set of dominant, male-specific values, such as rationality, control, and objectification. Thus, all reproductive technologies dominate and harm women and informed consent can be seen as a means of legalizing and thereby neutralizing this patriarchal and capitalist oppression. This context is empathy-poor because today, medical professionals are required to be value neutral, not to "feel with" or to put themselves in their patient's place, often contrary to the basic demands of the Hippocratic Oath.[27] In place of this affective bond they substitute a written contract. A contract is important but it is a poor substitute for empathy.

Empathy is covered up and marginalized by an overuse of reproductive technology and by an overreliance on medical contracts. Yet empathy is necessary for birth. For those who value birth, empathy can be seen as, what Marcuse calls a "vital need," a need that must be satisfied in order for the individual organism to thrive. This need ought to be addressed not just at the individual level but in terms of the institutions through which we develop socially valuable emotions, until, over time, they become a second nature. Marcuse's call for the satisfaction of the vital needs of freedom (Marcuse 1969, 10) and critical education, for example, can be expanded to include a woman's vital need for empathy in birth. Recognizing the value of empathy as vital to birth as a bringing-forth is the necessary first step toward institutionalizing this value. Addressing empathy as a woman's vital need in birth can allow it to "sink down"[28]—as Marcuse insightfully argues—into the "organic" behavior or structure of the medical institutions themselves, especially the practices of obstetricians and midwives. As Marcuse writes, "Once a specific morality is firmly established as a norm of social behavior, it is not only introjected—it also operates as a norm of 'organic' behavior: the organism receives and reacts to certain stimuli and 'ignores', repels others." (Marcuse 1969, 11).

Conclusion

Since women must still bear children in order for life on earth to continue, attention to childbirth matters. Who knows if women will choose to move further and further away from bringing children forth, preferring the challenging-forth alternative instead? I argue that the possibility of birth as a bringing-forth depends on the presence of empathy in childbirth. Recasting empathy as a vital need and a shared social value could begin to challenge the dominant and functional approach to childbirth that too often treats both mother and child as mere resources.

Notes

1. The gendered inverted dichotomy inherent in this distinction is interesting. The desire for rationality, control, planning, and order, which are typically associated with masculinity, are here (re)cast in Heidegger's concept of "leaping-in" and are most compatible with the reassuring medical fallback option. On the other hand, the desire and acceptance of contingency, irregularity, emotionality, and physical exposure to chaos and the unknown, which are traditionally associated with femininity, are privileged and (re)cast in Heidegger's concept of "leaping-ahead" and are more attuned to the less medicalized option.
2. John D. Caputo, *Radical Hermeneutics: Repetition, Deconstruction, and the Hermeneutic Project* (Bloomington: Indiana University Press, 1987), 189.

3. For more on the meaning of these and other birthing technologies, see Chapter 5.
4. Lauren Freeman, "Love Is Not Blind: In/Visibility and Recognition in Martin Heidegger's Thinking," *Institut für die Wissenschatften vom Menschen/Institute for the Human Sciences*, 107 (January 2013): quote on 2.
5. Ibid., 5.
6. For a detailed and sympathetic critique of the patriarchal and ethnocentric strands in Ruddick's account of maternal thinking see Jean Keller, "Rethinking Ruddick and the Ethnocentrism Critique of Maternal Thinking," *Hypatia* 25, no. 4 (2010): 834–851.
7. Ruddick appears to be drawing on Kant's "reflective judgment." This Kantian judgment proceeds from the particular to the universal where the universal is exemplified and discovered in particular experiences, rather than subsuming particular experiences under a universal umbrella, as in Kant's "determinant judgment."
8. See ch. 5.
9. Sara Ruddick, *Maternal Thinking* (Boston: Beacon, 1990), 193–195.
10. Ibid., 195.
11. Ibid., 209–210.
12. In her memoir, Isadora Duncan writes of pregnancy: "I began to be assailed with all sorts of fears... It was all in the course of life, etc. I was, nevertheless, conscious of fear. Of what? Certainly not of death, nor even of pain—some unknown fear, of what I did not know" (quoted in Simone de Beauvoir, *The Second Sex*, 504.
13. Ibid.
14. See Richard Rojcewicz, *The Gods and Technology: A Reading of Heidegger* (2006), 33.
15. Ibid., 24.
16. Is this empathetic attunement similar to what Heidegger in "Building Dwelling Thinking" describes as a saving, preserving, and dwelling?
17. For a detailed critique of the presumed ontological neutrality in Heidegger's *Being and Time* see Tina Chanter, "The Problematic Normative Assumptions of Heidegger's Ontology" in *Feminist Interpretations of Martin Heidegger*, ed. Nancy J. Holland and Patricia Huntington (Pennsylvania State University Press, 2001), 73–108.
18. This ambivalent relationship to technology, where one says both "yes" and "no" to technology is described by Heidegger in *Discourse on Thinking* (1966) as *Gelassenheit* or releasement toward things, and signals the preparatory stages in building a free relationship to the essence technology.
19. Freeman, "Love Is Not Blind," 4.
20. See http://abcnews.go.com/Health/utah-dad-posts-hospital-bill-40-fee-skin/story?id=42585001.

21. L. Shanner and J. Nisker, "Bioethics for Clinician: Assisted Reproductive Technologies," *Canadian Medical Association Journal* 164, no. 11 (2001): 1589–1594.
22. Ibid., 7. See also Inmaculada de Melo-Martin, "Ethics and Uncertainty: In Vitro Fertilization and the Risks to Women's Health," *Nine Risks: Health, Safety and Environment* (Summer 1998): 222.
23. de Melo-Martin, "Ethics and Uncertainty," 201. For instance, "if women are ignorant of the fact that IVF is of unproved benefit for many infertility conditions, and if they are not aware that the treatment may have unknown risks, then they cannot give genuinely informed consent." Since the official review of the procedure minimizes risks and maximizes success rates, the vast majority of women will not be fully informed and therefore their consent cannot be genuine or informed. Thus, according to one standard criticism of liberal feminism the latter tends to ignore the social conditions that shape the choices of individual women. For more on this view that also includes a discussion of the inevitability of social blind spots and thus partially restores a feminist merit to informed consent see Mary Anne Warren, "Is IVF Research a Threat to Women's Autonomy?" *Embryo Experimentation*, ed. Peter Singer, Helga Kuhse, Stephen Buckle, Karen Dawson, and Pascal Kasimba (Cambridge: Cambridge University Press, 1990), 125–140.
24. Robyn Rowland, "Choice, Control, And Issues of Informed Consent: The New Reproductive and Pre-Birth Technologies" (1986). For instance, "[liberal feminism] does not challenge social structures of inequality, nor the relations of production and reproduction. Through its stress on individual rights, liberalism places these rights in the private sphere...Though it is an important part of feminism, the personal alone is not enough...Individual women live out our lives within a social context. We are constrained and shaped by the forces of economics, social ideology, personal psychology, and the various power structures which mold our actions. There are social constraints operating upon choice, many of which are concerned with the social control of all persons but in this case of women. The world operates on the basis of power inequities. There is no equality in the alternatives offered to people as 'choices' and there is not equality between those who are 'choosing.' Decisions made between undesirable or negative alternatives hardly amount to free choice...Choices are impinged upon by ideological constructions, for example the pressure on women to be mothers...In addition the forces of capital and commerce blur for people a clear delineation of their needs as individuals and the needs which are socially constructed for us. People are expected and encouraged to choose socially acceptable alternatives. *This makes it very difficult for women during pregnancy and birth, for example, to resist the use of new technologies.* They can be accused of being selfish if they are not thinking of the child first. The 'maternal

consciousness' is shaped to be responsive to these arguments and to be responsive to science and medicine as 'problem-solvers'" (my emphasis). On a related but more moderate view of informed consent, see Mary Anne Warren, "Is IVF Research a Threat to Women's Autonomy?"

25. See Wendy McElroy's, "Feminists Against Women: The New Reproductive Technologies," at http://www.wendymcelroy.com. See also her *Liberty for Women: Freedom and Feminism in the Twenty-First Century* (Chicago: Ivan R. Dee and the Independent Institute, 2002).

26. Robyn Rowland, "Choice, Control, and Issues of Informed Consent: The New Reproductive and Pre-Birth Technologies" (1986). Inmaculada de Melo-Martin underscores this point in "Ethics and Uncertainty: In Vitro Fertilization and the Risks to Women's Health," *Nine Risks: Health, Safety and Environment* (Summer 1998): 84–122. She shows how the legal evaluation of the risks and side effects of IVF by international government IVF commissions consistently undermine risks and uncertainties associated with the procedure. Making a procedure seem safer than it actually is or than it is known to be undermines a woman's ability to provide genuine informed consent but does not undermine the appeal of informed consent.

27. The modern formulation of the Hippocratic Oath requires that the doctor is sympathetic with the patient and that she regard the patient as a whole person and not reduce the patient to his or her symptom. These requirements, ignored by informed consent, are in line with an empathetic approach.

28. Marcuse writes "If biological needs are defined as those which must be satisfied... certain cultural needs can 'sink down' into the biology of man. We could then speak, for example, of the biological need of freedom (and of empathy), or of some aesthetic needs as having taken root in the organic structure of man, in his 'nature,' or rather 'second nature.'" See Herbert Marcuse, *An Essay on Liberation* (Boston: Beacon, 1969), 10–11, my insert.

Epilogue: Heidegger's *Black Notebooks*

While I was revising this manuscript, the first volumes of Heidegger's *Black Notebooks* (*Schwarze Hefte/GA 94–96*) were published in Germany and were publicly excoriated. It is beyond the scope of my present work to comment on the various failings of the *Black Notebooks*. It is relevant to note, however, that GA 94–96, spanning the years 1931–1941, raise important hermeneutical questions and concerns about machination (*Machenschaft*) as the genesis of Heidegger's later concept of *Ge-stell* or enframing. Consistent with the ontologization of machination and enframing in GA 65, GA 5, GA 7, and GA 79, machination in GA 94–96, is described as a principle that organizes entire Western societies; a sending of being, an *epoch(é)* of our forgetfulness of being. According to Heidegger, "the highest reign of Being as machination spreads the complete forgetfulness of Being all around" (GA 95, 385; my translation). Machination hides (94: 296) in the age of the domination of the lack of questioning of Being (94: 315) and it takes shape as: science, technology, culture" (GA 94: 420).

The entire Western world, as Heidegger sees it in GA 94–96, is caught up in machination. Communism, Bolshevism, Socialism, Fascism, National Socialism, Christianity, Judaism, Americanism, and the English as champions of utilitarianism (GA 96: 109, 111; GA 94: 432; GA 95: 188; GA 95: 408; GA 96: 266; GA 96: 263) all promote the homogenization of thinking, the mechanization of nature and of human relationships, and the leveling of all meaningful differences.

Heidegger's use of machination in 1934 to sum up these different peoples, politics, and religions is recklessly abstract and an instance of the leveling that he criticizes. While abstraction persists in the later concept of the enframing, the tendency there is more restricted to analyzing the triangulation between technological use, the technologization of human thought and the possibility of human freedom in the technical world. Although problematic, the proneness toward abstraction is tempered by rich phenomenological descriptions of the technicization of nature that continue to inspire readers to this day. Clearly any analysis that casts Bolshevism and National Socialism, from 1931 to 1941, *merely* as forms of machination is deeply problematic and mystified. Bombastic claims that "the new politics is born out of the essence of technology" (94: 472), along with his belief in the greatness of National Socialism despite its current form (95: 408), attest to this mystification and reveal his anti-Semitism, while shedding little light on machination as the essence of technology (GA 95: 392). Since machination is so poorly explained, it cannot explain anything about the politics of Bolshevism or National Socialism. Worse, since machination is described primarily as a sending of being, a symptom of our forgetfulness of being that is *a fortiori* an impersonal legacy, it annuls any responsibility for the crimes of Bolshevism and National Socialism. Since these crimes were already documented by 1941, this shirking of responsibility, along with Heidegger's continued membership in the Nazi party, are inexcusable. Worse still, the repeated causal associations between machination and World Jewry are racist and condemnable. Moreover, they are philosophically incoherent.

On Heidegger's own account, machination cannot both be a sending of being and brought on by World Jewry, and yet his writings in GA 94–96 sometimes equivocate between these two origins. In other words, machination cannot have two origins, one ontological and the other ontic. If World Jewry is the bearer of machination in some primordial and world-historical (*geschichtlich*) way, Heidegger's writings do not support this interpretation. Heidegger's implications that the "worldless" and "calculating" (GA 95: 97) World Jewry has an especial affinity for enframing and for the growing danger in the West are marks of his anti-Semitism in 1931–1941, but find no support in his 1949–1954 philosophical writings on the enframing. While these implications are shameful, they do not seem especially relevant to a critical application of enframing as a phenomenological principle, used to reveal the recasting of women's reproductive bodies as technological resources. His politics would be relevant only to the extent that it could be shown that anti-Semitic (or anti-Bolshevik or anti-Christian) empirical considerations insinuate themselves into the structure of enframing.

Belu, D.S.

BIBLIOGRAPHY

Aken, H., and W. Gogarten (2000) "A Century of Birth" in *The International Anesthesia Society*.
Arditti, R., R. Duelin Klein, and S. Minden (1989) *Made to Order: The Myth of Reproductive and Genetic Progress* (London: Pandora Press).
Arendt, H. (1958) *The Human Condition* (Chicago: The University of Chicago Press).
Arendt, H. (1964) *Eichmann in Jerusalem: A Report on the Banality of Evil*, Revised and Enlarged Edition. (New York: Viking Press).
Aristotle (1941) *The Basic Works of Aristotle* ed. McKeon, Richard (New York: Random House).
Arms, S. (1994) *Immaculate Deception II* (Berkeley: Celestial Arts Publishers).
Bailey, A. (2011) "Reconceiving Surrogacy: Toward a Reproductive Justice Account of Indian Surrogacy" in *Hypatia*, vol. 26, no. 4, 715–741.
Banerjee, A. (2011) *Reconceiving "Borders": A Feminist Pragmatic Phenomenology for Postcolonial Ethics and Politics*. Dissertation. University of Oregon.
Bartky, S. (1990) *Femininity and Domination: Studies in the Phenomenology of Oppression* (New York: Routledge Press).
Beauvoir, S. (1989) *The Second Sex* trans. H.M. Parshley (New York: Vintage Books).
Bigwood, C. (1995) *Earth Muse: Feminism, Nature, and Art* (Philadelphia: Temple University Press).
Buckle, S., K. Dawson, H. Kuhse, P. Kasimba, and P. Singer eds. (1992) *Embryo Experimentation: Ethical, Legal and Social Issues* (New York: Cambridge University Press).

Burke, C, N. Schor, and M. Whiford eds. (1994) *Engaging with Irigaray: Feminist Philosophy and Modern European Thought* (New York: Columbia University Press).

Burrow, S. (2012) "On the Cutting Edge: Ethical Responsiveness to Rising Cesarean Rates" in *American Journal of Bioethics*, vol. 12, no. 5, 44–52.

Cartwright. E. (1998) "The Logic of Heartbeats: Electronic Fetal Monitoring and Biomedically Constructed Birth" in *Cyborg Babies from Techno-Sex to Techno Tots*, eds. R. Davis-Floyd & J. Dumit (New York: Routledge Press).

Caputo, J.D. (1987) *Radical Hermeneutics: Repetition, Deconstruction, and the Hermeneutic Project* (Bloomington: Indiana University Press).

Caputo, J.D. (1993) "*Aletheia* and the Myth of Being" in *Demythologizing Heidegger* (Bloomington: Indiana University Press).

Corea, G. (1986) *The Mother Machine: Reproductive Technologies from Artificial Insemination to Artificial Wombs* (New York: HarperCollins Pubs.).

Davis-Floyd, R. (1998) *Cyborg Babies: From Techno-Sex to Techno Tots* (New York: Routledge Press).

Derrida, J. (1997) *Of Grammatology*, trans. Gayatri Chakrovorty Spivak (Baltimore: Johns Hopkins University Press).

Dreyfus, H. (1995) "Heidegger on Gaining a Free Relation to Technology" in *Technology & The Politics of Knowledge*, eds. A. Feenberg and A. Hannay (Bloomington: Indiana University Press).

Feenberg, A. (1999) *Questioning Technology* (New York: Routledge Press).

Feenberg, A. (2000) "The Ontic and the Ontological in Heidegger's Philosophy of Technology: Response to Thomson" in *Inquiry*, vol. 43, 445–450.

Feenberg, A. (2005) *Heidegger and Marcuse on the Catastrophe and Redemption of Modernity* (New York: Routledge Press).

Ferrell, R. (2006) *Copula: Sexual Technologies, Reproductive Powers* (Albany: State University of New York Press).

Firestone, S. (1970) *The Dialectic of Sex: The Case for a Feminist Revolution* (New York: Farrar Strauss & Giroux).

Foucault, M. (1977) *Discipline and Punish: The Birth of the Prison* trans. A. Sheridan (New York: Vintage Books)

Foucault, M. (1994) *The Order of Things: An Archaeology of the Human Sciences* (New York: Vintage Press).

Freeman, L. (2013) "Love is Not Blind: In/Visibility and Recognition in Martin Heidegger's Thinking," *Institut für die Wissenschaften vom Menschen/Institute for the Human Sciences*. www.iwm.at.

Freud, S. (1961) *Civilization and its Discontents* trans. J. Strachey (New York: Norton Books).

Garry, A. (2001) "Medicine and Medicalization: A Response to Purdy" in *Bioethics*, vol. 15, no. 3, 262–269.

Glazebrook, T. (2000) "From φύσις to Nature, τέχνη to Technology: Heidegger on Aristotle, Galileo and Newton" in *The Southern Journal of Philosophy*, vol. XXXVIII, 95–118.
Greenspan, D. (2008) *The Passion of Infinity: Kierkegaard, Aristotle and the Rebirth of Tragedy* (New York: Walter DeGruyter Press).
Guenther, L. (2006) *The Gift of the Other: Levinas and the Politics of Reproduction* (Albany: State University of New York Press).
Harding, S. (1991) *Whose Science? Whose Knowledge? Thinking from Women's Lives* (New York: Cornell University Press.).
Harwood, K. (2007) *The Infertility Treadmill: Feminist Ethics, Personal Choice, and the Use of Reproductive Technologies* (University of North Carolina Press).
Heidegger, M. (1949) *Gesamtausgabe Band 79* (Frankfurt: Vittorio Klosterman).
Heidegger, M. (1952) *Feldweg* (Frankfurt: Vittorio Klosterman).
Heidegger, M. (1954) *Vorträge und Aufsätze* (Pfüllingen: Günther Neske Verlag).
Heidegger, M. (1954) *The Question Concerning Technology and Other Essays*, trans. W. Lovitt (New York: Haper & Row Publishers).
Heidegger, M. (1958) *The Question of Being* trans. William Kluback and Jean T. Wilde (Twayne Publishers).
Heidegger, M. (1962/1998) "Traditional Language, Technological Language," trans. W. T. Gregory, *Journal of Philosophical Research*, vol. 23, 129–145.
Heidegger, M. (1966) *Discourse on Thinking* trans. John M. Anderson (New York: Harper & Row Publishers.)
Heidegger, M. (1971) *Poetry, Language, Thought* ed. J. Glenn Gray trans. A. Hofstadter (New York: Harper & Row Publishers).
Heidegger, M. (1975) "The Way Back into the Ground of Metaphysics" in *Existentialism from Dostoevsky to Sartre*, ed. & trans. Walter Kaufmann (New York: Meridian Press).
Heidegger, M. (1976) "Only a God Can Save Us" in *Der Spiegel Interview with Martin Heidegger* trans. John. D. Caputo and Maria P. Alter *Philosophy Today*, XX, 267–284.
Heidegger, M. (1977) *The Question Concerning Technology and Other Essays* trans. W. Lovitt (New York: Harper & Row, Publishers, Inc.).
Heidegger, M. (1996) *Being and Time* trans. J. Stambaugh (Albany: State University of New York Press).
Heidegger, M. (1998) "On the Essence and Concept of φύσις in Aristotle's Pysics B, I" in *Pathmarks*, ed. W. McNeill. (New York: Cambridge University Press).
Heidegger, M. (1999a) *Contributions to Philosophy (From Enowning)* trans. P. Emad and K. Maly (Bloomington: Indiana University Press).
Heidegger, M. (1999b) *Pathmarks* W. McNeill (New York: Cambridge University Press).
Heidegger, M. (2003) *The End of Philosophy*, trans. J. Stambaugh (Chicago: The University of Chicago Press).

Heidegger, M. (2012) *Bremen and Freiburg Lectures: Insight into That Which Is and Basic Principles of Thinking* trans. A. Mitchell (Bloomington: Indiana University Press).

Heidegger, M. (2014) *Gesamtausgabe Bände 94–96, 1934–1941 Schwarze Hefte 1934–1941* herausgegeben von P. Trawny (Frankfurt am Main: Vittorio Klosterman).

Held, V. (1989) "Birth and Death" in *Ethics*, vol. 99, no. 2, 362–388.

Holland, N., and P. Huntington eds. (2001) *Feminist Interpretations of Martin Heidegger* (University Park: The Pennsylvania State University Press).

Husserl, E. (1989) *Ideas Pertaining to a Pure Phenomenology and to a Phenomenological Philosophy* trans. R. Rojcewicz and A. Schuwer (Dodrecht: Kluwer Academic).

Ihde, D. (2010) *Heidegger's Technologies: Postpphenomenological Perspectives* (New York: Fordham University Press).

Irigaray, L. (1993) *An Ethics of Sexual Difference* trans. C. Burke and G.C. Gillian (New York: Cornell University Press).

Irigaray, L. (1999) *The Forgetting of Air in Martin Heidegger* trans Mary Beth Mader (Austin: University of Texas Press).

Jonas, H. (2012) "Technology and Responsibility: Reflections on the New Task of Ethics" in *Society, Ethics, and Technology*, 4th ed., eds. M. Winston and R. Edelbach (Wadsworth Press).

Katz-Rothman, B. (2000) *Recreating Motherhood* (New Brunswick: Rutgers University Press).

Keller, J. (2010) "Rethinking Ruddick and the Ethnocentrism Critique of Maternal Thinking" in *Hypatia*, vol. 25, no. 4, 834–851.

Kierkegaard, S. (1987) *Either/Or (Part I)* trans. H. V. Hong and E. H. Hong (Princeton: Princeton University Press).

Kisiel, T. (2016) *Heidegger, History and the Holocaust* by Mahon O'Brien, Reviewed in *Notre Dame Philosophical Reviews*, http://ndpr.nd.edu/.

Klawiter, M. (1990) "Using Arendt and Heidegger to Consider Feminist Thinking on Women and Reproductive/Infertility Technology" in *Hypatia*, vol. 5, no. 3, 69–81.

Laborie, F. (1987) "Looking for Mothers You Only Find Fetuses" in *Made to Order: The Myth of Reproductive and Genetic Progress*, eds. P. Steinberg & D. Lynn (Oxford: Pergamon Press).

LaChance Adams, S., and C. Lundquist (2013) *Coming to Life: Philosophies of Pregnancy, Childbirth, and Mothering* (New York: Fordham University Press).

Lamaze, F. (1970) *Painless Childbirth: Psychoprophylactic Method* trans. L.R. Celestin (Chicago: Henry Regnery Co.).

Lee, W. L. (2014) *Contemporary Feminist Theory and Activism: Six Global Issues* (New York: Broadview Press).

Levinas, E. (1969) *Totality and Infinity* trans. A. Lingis (Pittsburgh: Duquesne Press).
Longino, H. (1995) "Knowledge, Bodies, and Values: Reproductive Technologies and Their Scientific Context," in *Technology & The Politics of Knowledge*, ed. A. Feenberg & A. Hannay (Bloomington: Indiana University Press), 195–210.
Loudon, I. (2002) *Death in Childbirth: An International Study of Maternal Care and Maternal Mortality 1800–1950* (Oxford: Clarendon Press).
Marcuse, H. (1964) *One-Dimensional Man: Studies in the Ideology of Advanced Industrial Society* (Boston: Beacon Press).
Marcuse, H. (1969) *An Essay on Liberation* (Boston: Beacon Press).
Markotic, L. (2016) "Paternity, Enframing and a New Revealing: O'Brien's Philosophy of Reproduction and Heidegger's Critique of Technology," *Hypatia*, vol. 31, no. 1.
McElroy, W. (2002) *Liberty for Women: Freedom and Feminism in the Twenty First Century* (Chicago: Ivan R. Dee & The Independent Insitute).
Melo-Martin, I.M. (1998) "Ethics and Uncertainty: In Vitro Fertilization and the Risks to Women's Health" in *9 Risks: Health, Safety and Environment*, vol. 9, no. 3, 201–227.
Merchant, C. (1989) *The Death of Nature: Women, Ecology and the Scientific Revolution* (San Francisco: Harper & Row, Publishers).
Merleau-Ponty, M (1968) *The Visible and the Invisible* trans. A. Lingis (Chicago: Northwestern University Press).
Michaels, P. (2014) *Lamaze: An International History* (New York: Oxford University Press).
Minden, S. (1984) "Patriarchal Designs: The Genetic Engineering of Human Embryos" in *Test Tube Women*, eds. R. Arditti, R. Klein, and S. Minden (London: Pandora Press).
Moi, T. (1986) *The Kristeva Reader* (New York: Columbia University Press).
Nussbaum, M. (2000) *Sex & Social Justice* (New York: Oxford University Press).
Oliver, K. (2010) "Motherhood, Sexuality and Pregnant Embodiment: Twenty-Five Years of Gestation" in *Hypatia*, vol. 25, no. 4, 760–777.
Oliver, K. (2013) *Technologies of Life and Death: From Cloning to Capital Punishment* (New York: Fordham University Press).
Pande, A. (2011) "Transnational Surrogacy in India: Gift for Global Sister?" in *Reproductive Biomedicine Online*, vol. 23, 618–625.
Pande, A. (2014) "This Birth and That: Surrogacy and Stratified Motherhood in India" in *philo SOPHIA*, vol. 4, no. 1, 50–64.
Panitch, V. (2013) "Global Surrogacy: Exploitation to Empowerment," *Journal of Global Ethics*, vol. 9, no. 3, 329–343.

Raymond, J. (1993) *Women as Wombs: Reproductive Technologies and the Battle Over Women's Freedom* (San Francisco: Harper Press).

Rich, Adrienne (1986) *Of Woman Born: Motherhood as Experience and Institution* (New York: W.W. Norton & Co).

Roberts, Dorothy (1999) *Killing the Black Body: Race, Reproduction and the Meaning of Liberty* (New York: Vintage Books).

Rodin, J., and A. Collins (1991) *Women and New Reproductive Technologies: Medical, Psychosocial, Legal and Ethical Dilemmas* (Laurence Erlbaum Associates, Inc.).

Rojczewicz, R. (2006) *The Gods and Technology: A Reading of Heidegger* (Albany: State University of New York Press).

Rowland, R. (1986) "Choice, Control, and Issues of Informed Consent: The New Reproductive And Pre-Birth Technologies." www.finrage.org.

Ruddick, S. (1990) *Maternal Thinking: Toward a Politics of Peace* (Boston: Beacon Press).

Ruin, H. (2012) "Technology as Destiny, in Heidegger and Cassirer" in *Cassirer on Form and Technology*, ed. A. Hoe (London: Palgrave), 133–156.

Sandel, M. (2012) "The Case Against Perfection" in *Society, Ethics, and Technology*, 4th ed., eds. M. Winston and R. Edelbach (Wadsworth Press).

Saravanan, S. (2013) "An Ethnomethodological Approach to Examine Exploitation in the Context of Capacity, Trust and Experience of Commercial Surrogacy in India" in *Philosophy Ethics and Humanities in Medicine, open access https://peh-med.biomedcentral.com/articles/10.1186/1747-5341-8-10#Sec1, 2–12.*

Schummer, J. (2001) "Aristotle on Technology and Nature" in *Philosophia Naturalis*, vol. 38, 105–120.

Sheehan, T. (2015) *Making Sense of Heidegger: A Paradigm Shift* (New York: Rowman & Littlefield International).

Sher, G., V. Davies, and J. Stoes (1995) *In Vitro Fertilization: The A.R.T. of Making Babies* (New York: Facts on File).

Singer, P., H. Kuhse, S. Buckle, K. Dawson, and P. Kasimba eds. (1990) *Embryo Experimentation* (New York: Cambridge University Press).

Smerdon, U. (2009) "Crossing Bodies, Crossing Borders: International Surrogacy between The United States and India" in *Cumberland Law Review*, vol. 39, 15–85.

Sophocles (1982) *The Complete Plays of Sophocles* Sir Richard Claverhouse Jebb (New York: Bantam Books).

Spallone, P., and D.L. Steinberg (1987) *Made to Order: The Myth of Reproductive and Genetic Progress* (New York: Pergamon Press).

Spengler, O. (1932) *Man and Technics: A Contribution to a Philosophy of Life* (New York: Alfred Knopf Publishers).

Teman, E. (2003) "The Medicalization of 'Nature' in the 'Artificial' Body: Surrogate Motherhood in Israel" in *Medical Anthropology Quarterly*, vol. 17, no. 1, 78–98.
Thatcher, S., and A. DeCherney (1991) "Pregnancy-Inducing Technologies: Medical, Psychosocial, Legal and Ethical Dilemmas" in *Women & New Reproductive Technologies: Medical, Psychosocial, Legal and Ethical Dilemmas* (Lawrence Erlbaum Associate, Inc.).
Thomson, I. (2005) *Heidegger on Ontotheology: Technology and the Politics of Education* (New York: Cambridge University Press).
Verbeek, P-P. (2005) *What Things Do: Philosophical Reflections on Technology, Agency, and Design* (University Park: The Pennsylvania State University Press).
Verny, T. (1981) *The Secret Life of the Unborn Child* (New York: Bantam Doubleday Dell Publishing Group).
Vora, K. (2013) "Potential Risk and Return in Transnational Indian Surrogacy" in *Current Anthropology*, vol. 54, no. S7, 97–106.
Wajcman, J. (1991) *Feminism Confronts Technology* (University Park: Pennsylvania University Press).
Wilkinson, S. (2010) *Choosing Tomorrow's Children: The Ethics of Selective Reproduction* (New York: Oxford University Press).
Wolin, Richard (1993) *The Heidegger Controversy: A Critical Reader* (Cambridge, MA: MIT Press).
Wolin, R., and J. Abromeit eds. (2005) *Heideggerian Marxism: Marcuse, Herbert* (Lincoln: University of Nebraska Press).
Young, I.M. (2005) *On Female Body Experience: Throwing like a Girl & Other Essays* (New York: Oxford University Press).
Zwelling, E. (2001) "The History of Lamaze Continues: An Interview with Sunnyie Strickland" *The Journal of Perinatal Education*, Winter vol. 10, 15–21.

Index

A
Abetting, 113
Advanced reproductive technology (ART), 111
Agency, 6, 11, 40, 50, 68, 79, 87, 95, 96
  See also Maternal agency; *Subversive agency*
Aletheia, 2, 10, 14, 18n6
Arendt, Hannah, 3, 38, 49, 56n53, 81
Aristotle, 2, 3, 5, 61–69, 71, 73n1, 73n2, 74n3, 74n4, 75n12
Artificial conception, 4, 5, 17, 23–52, 61–73
Artificial womb, 35, 54n38
Attentive love, 104, 108, 109, 114
Authenticity, 108, 109
  See also Inauthenticity

B
Bailey, Alison, 45, 46, 48, 50
Beauvoir, Simone de, 112, 117n12
Bigwood, Carol, 64, 73n2, 74n3, 111
Biological ownership, 72
Birthing body, 110, 111
Body, 28, 34–37, 39, 45–48, 50, 52n6, 57n60, 59n73, 66, 73, 78, 85–89, 91, 94, 95, 97, 98n4, 100n21, 101n27, 101n30, 110, 113
  See also Birthing body; Lived body; Object-body
Bringing-forth, 6, 81, 104, 110–114, 116

C
Calculative thinking, 4, 24, 56n52
Causality, 69, 71
  See also Efficient cause; Final cause; Formal cause; Material cause
Cesarean section, 2, 88, 89
Challenging-forth, 3, 19n7, 22n35, 24, 69–71, 78, 79, 111, 116
Childbirth, 5, 17, 24, 26, 77, 78, 81–86, 89–97, 98n9, 98n10, 100n26, 104, 106, 108–113, 116
  See also Natural childbirth; Technological childbirth
Clearing, 4, 9, 11, 13, 16, 104
Concealing, 2, 111
Conception, 1–5, 52n7, 56n52, 59n80, 75n12, 95, 98n9, 100n19, 111
  See also Artificial conception

D
Decontextualization, 30, 31, 32, 35, 44

E

Ectogenesis, 35
Efficiency, 5, 38, 78, 79, 91, 95, 96, 99n11, 100n23
Efficient cause, 68, 69, 71, 75n12, 113
Empathy, 6, 27, 104, 105, 109, 110, 113–116, 119n28
Enframed, 3, 4, 9, 11–16, 22n32, 38, 93, 104
Enframed thinking, 38
Enframing, 1–6, 8–17, 18n6, 19n11, 21n26, 23–52, 55n46, 56n55, 61, 69, 71, 73, 78–81, 85, 88, 89, 91–94, 96, 97, 103–105, 109
See also Enframed; Ge-stell; Reproductive enframing
Ethics, 32, 33, 45, 52n5, 57n58, 63, 98n10, 109
Ethnography, 44, 45, 48, 50
Existential, 22n30, 113

F

Fear, 81, 94, 110, 111, 112, 117n12
Feenberg, Andrew, 3, 20n12, 29–33, 70
Feminism, 1–6, 8, 17, 25, 26, 28, 34, 36, 39, 40, 44, 45, 47–50, 58n66, 72, 81, 82, 89, 91, 92, 95, 96, 97, 98n9, 100n20, 101n30, 104, 107, 110, 114, 115, 118n23, 118n24
See also Feminist philosophy; Liberal feminism; Radical feminism
Feminist phenomenology, 3, 8, 17, 25, 28, 44, 52n5, 81, 97, 104
Feminist philosophy, 3
Fertility, 27, 29, 34, 36, 54n22, 60n81, 62, 65–68, 71, 72
See also Fertility doctor; Fertilization
Fertility doctor, 62, 66, 67, 68, 71, 72, 75n14

Fertilization, 27, 31, 33, 35, 36, 52n6, 66
Fetal distress, 89
Fetus, 36, 45–48, 50, 67, 86, 87, 89–91, 99n16, 104, 109, 111–114
See also Fetal distress
Final cause, 71
Formal cause, 71
Fragmentation, 4, 27, 28, 31, 35, 37, 51
Freedom, 11, 27, 34, 38, 111, 115, 119n28, 122
Functionality, 29–36, 40, 41, 44, 77, 114, 116
Fungibility, 9, 30, 36, 37, 46, 47, 49, 71, 89

G

Gestational surrogacy, 4, 25, 26, 39–41, 44, 45, 48, 50, 51, 58n66, 58n68
Ge-stell, 8, 40, 50, 78
Glazebrook, Trish, 64, 66, 74n3

H

Heidegger, Martin, 1–6, 7–17, 19n10, 20n13, 21n25, 21n28, 22n30, 22n32, 22n35, 23–25, 28, 29, 36, 37, 40–43, 48–51, 55n40, 55n46, 56n52, 56n55, 57n60, 61, 62, 67, 69–71, 73, 78–85, 91, 97, 97n1, 98n9, 100n19, 103–107, 110, 112–114, 116n1, 117n18
Historicity, 2, 4, 5, 8, 11, 13, 14, 16, 24, 41, 57n58, 68, 78, 79, 80, 90
History, 2, 4, 9, 10, 13, 18n3, 20n13, 31, 34, 82, 83, 91, 97n1, 111
See also Historicity
History of being, 2, 9, 18n3, 56n55, 97n1

I

Imitation, 32, 65, 66, 70, 74n4
Inauthenticity, 105, 106, 107
Informed consent, 6, 114, 115, 118n23, 118n24, 119n26, 119n27
Instrumentalism, 79
  See also Instrumental reason; Instrumental thinking
Instrumentalization Theory, 29, 33
  See also Primary instrumentalization; Secondary instrumentalization
Instrumental reason, 24, 37, 38
Instrumental thinking, 38, 39, 56n53
In vitro fertilization, 4, 25, 52n6, 62, 118n22, 119n26
  See also IVF
IVF, 4, 5, 25–36, 38, 39, 41, 52n6, 53n9, 53n12, 53n16, 62, 63, 65–73, 75n12, 75n14, 101n30, 118n23, 119n26

K

Knowledge, 14, 15, 16, 17, 22n32, 52n6, 77, 82, 94, 96
  See also Prereflective knowledge; Representational knowledge

L

Lamaze, Fernand, 3, 5, 6, 78, 91–97, 100n24, 101n27, 101n28, 101n29, 109, 110, 114
Leaping
  leaping-ahead, 104, 105, 116n1
  leaping-in, 104–106, 109, 114, 116n1
Liberal feminism, 6, 25, 44, 115, 118n23, 118n24
Lifeworld, 9, 29, 30, 31, 32, 44
Lived body, 58n69
Louden, Irvine, 82

M

Marcuse, Herbert, 55n46, 55n47, 56n52, 115, 116, 119n28
Markotic, Lorraine, 16, 22n33, 81, 98n8
Material cause, 64, 75n12
Maternal agency, 104–110
Maternal care, 40
Maternal thinking, 6, 104, 106, 108, 117n6
  See also Maternal care; Maternal work
Maternal work, 41, 51, 60n81
Medicalization, 3, 6, 25, 28, 33, 47, 68, 83, 89, 92, 93, 98n10, 100n23, 100n24, 105, 107, 109, 111, 113, 116n1
Meditative thinking, 15, 22n30, 22n31
  See also Releasement toward things
Metaphysics, 2, 15, 64, 97n1
Michaels, Paula, 82, 86, 88, 93, 95, 101n29
Midwife, 82, 83, 92, 100n23, 100n26, 114
The mother-effect, 33, 51, 52, 60n83, 72, 73
Motherhood, 1, 4, 32, 39, 40, 42, 45, 47, 50–52, 70, 83

N

Natural childbirth, 77, 86
Nature, 5, 8, 10, 12–15, 18n6, 19n11, 33–37, 40, 43, 47, 48, 51, 55n40, 59n73, 60n83, 61–73, 73n1, 74n3, 74n4, 75n12, 78–80, 92, 94–96, 98n4, 109, 112, 115, 119n28
  See also Physis
Network, 5, 17n1, 27–29, 32, 44, 45, 47, 49, 50, 85, 87–89, 91, 100n19

O

Object, 5, 6, 15, 19n11, 24, 26, 30–32, 36, 37, 41, 42–44, 47–51, 56n53, 57n63, 63, 69, 79, 81, 85–88, 90–92, 95, 96
See also Object-body; Objectification
Object-body, 46
Objectification, 5, 37, 43, 50, 100n22, 108, 115
Oliver, Kelly, 33, 35, 40, 51, 59n74, 60n83, 72
Ontic, 9, 11, 57n61, 122
Ontology, 8, 16, 55n46, 68, 104
Optimization, 3, 5, 8, 28, 39, 43, 62, 78, 79, 92, 96, 97, 100n19
Order(ing), 9, 10, 11, 14, 27–29, 31, 33, 35, 39, 40, 42–44, 48–50, 52n6, 61, 62, 70, 78–80, 82, 86, 87, 89–92, 94, 95, 97, 99n11, 105, 115, 116n1

P

Pande, Amrita, 44, 45, 48
Paradox, 4, 9, 13, 16, 17, 22n32, 72, 103
Phenomenology, 2, 3, 17, 78, 81, 98n10, 110, 112
See also Feminist phenomenology
Physis, 5, 61, 64, 74n3, 81, 98n9, 112
Poiesis, 6, 81, 104, 112
Pregnancy, 25, 26, 28, 33, 35, 36, 45, 47, 48, 53n12, 55n38, 59n73, 67, 70, 90, 92, 100n19, 112, 117n12
Prereflective knowledge, 24, 72
Primary instrumentalization, 30, 32, 33
Primary qualities, 31
Psychoanalytic, 59n73, 60n83, 72
Psychoprophylactic, 93

Q

Qualities, 31, 32, 35, 45
See also Primary qualities; Secondary qualities

R

Racism, 56n53, 58n68
See also Racist ideology
Racist ideology, 91
Radical feminism, 6, 26, 34, 47, 58n71, 89, 100n20, 101n30, 115
Rationality, 55n46, 56n52, 65, 79, 110, 115, 116n1
Raw material, 3, 4, 9, 11, 12, 24, 71, 78
Reason, 11, 16, 24, 37, 38, 53n12, 57, 67, 79, 83, 89, 96, 110, 111, 112
See also Instrumental reason; Rationality
Reification, 56n52
Releasement toward things, 22n31, 117n18
Representational knowledge, 22n30, 110
Reproductive enframing, 3–5, 24, 25, 28, 32, 34, 35, 37, 38, 40, 50, 51, 73, 88, 89, 92–94, 97, 99n11, 100n19, 100n23, 104, 105, 109
Resource, 3–6, 8, 11, 13, 14, 19n11, 24, 28–30, 33, 37–45, 47–51, 57n62, 57n63, 59n72, 73n2, 78–81, 85, 87–89, 99n19, 104
Resource thinking, 37–38
Revealing, 2, 8, 9, 13, 14, 15, 16, 21n28, 24, 41, 71, 78
See also Concealing
Rich, Adrienne, 3, 83, 84, 85, 100n21, 113
Roberts, Dorothy E., 34, 39, 90, 91, 98n10
Rojcewicz, Richard, 113

Rothman, Katz Barbara, 56n54, 56n55, 83, 89, 90
Ruddick, Sara, 3, 6, 104, 105–111, 113, 114, 117n6, 117n7

S

Schummer, Joachim, 65, 74n5, 75n12
Secondary instrumentalization, 29, 30, 32, 33
Secondary qualities, 31, 32
Sheehan, Thomas, 9, 55n40, 57n59
Spare space, 46, 48
Splitting, 48, 59n73
Standing-reserve, 3, 10, 11, 13, 14, 19n11, 20n14, 24, 36, 42, 51, 69, 79
  See also Resource
Subject, 6, 9, 13, 24, 28, 29, 32, 35, 36, 37, 42, 43, 44, 48–51, 56n52, 57n58, 57n63, 59n73, 79, 81, 85, 87, 90–92, 94, 95, 105
  See also Subjectification; Subjectivism; Subjectivity
Subjectification, 33, 34, 108
Subjectivism, 79
Subjectivity, 29, 50
Subversive agency, 45, 48
Superovulation, 27, 31, 66
Surrogacy, 4, 25, 26, 39, 40, 41, 43–45, 47–51, 58n68, 98n9
  See also Gestational surrogacy; Transnational surrogacy

T

Techne, 5, 35, 58n66, 61, 62, 64–68, 71, 74n3, 74n4, 82, 94, 95

Technological childbirth, 78
Technological thinking, 14, 38, 39, 56n53
  See also Enframed thinking; Resource thinking
Technology, 2–5, 7–11, 13, 15, 17, 18n2, 18n6, 21n26, 23–26, 28, 30, 32–35, 37, 48, 50, 51, 55n46, 62–63, 65, 69–73, 78–84, 88, 91, 93, 94, 97, 100n23, 101n30, 105, 110, 111, 113, 114, 115, 117n18
  See also Advanced reproductive technology (ART); Techné; Technophilia; Technophobia
Technophilia, 77, 80, 81
Technophobia, 77, 81
Telos, 65, 71, 74n6
Throwness, 81, 110
Transnational surrogacy, 40, 50

V

Vora, Kalindi, 44, 45, 46, 48, 49, 59n72

W

Water birth, 104, 112, 113, 114
Womb, 4, 5, 24, 27, 28, 31, 32, 35, 46, 61, 66, 67, 71, 101n30, 111
Work of conscience, 104, 107–110
World, 2, 3, 4, 9, 11–13, 15, 24, 39, 41–46, 50, 70, 78, 82, 96, 110, 112, 114, 118n24
  See also Worldlessness
Worldlessness, 41

The manufacturer's authorised representative in the EU is Springer Nature Customer Service Centre GmbH, Europaplatz 3, 69115 Heidelberg, Germany. If you have any concerns regarding our products, please contact ProductSafety@springernature.com

Printed and bound by CPI Group (UK) Ltd, Croydon, CR0 4YY

23/03/2026

02076402-0004